Joseph Kirtland

Complementation of Normal Subgroups

Also of Interest

Simple Lie Algebras over Fields of Positive Characteristic. Volume I: Structure Theory
Helmut Strade, 2017
ISBN 978-3-11-051516-9, e-ISBN (PDF) 978-3-11-051544-2,
e-ISBN (EPUB) 978-3-11-051523-7

Simple Lie Algebras over Fields of Positive Characteristic. Volume II: Classifying the Absolute Toral Rank Two Case
Helmut Strade, 2017
ISBN 978-3-11-051676-0, e-ISBN (PDF) 978-3-11-051760-6,
e-ISBN (EPUB) 978-3-11-051689-0

Groups of Prime Power Order. Volume 4
Yakov G. Berkovich, Zvonimir Janko, 2016
ISBN 978-3-11-028145-3, e-ISBN (PDF) 978-3-11-028147-7,
e-ISBN (EPUB) 978-3-11-038155-9

Groups of Prime Power Order. Volume 5
Yakov G. Berkovich, Zvonimir Janko, 2016
ISBN 978-3-11-029534-4, e-ISBN (PDF) 978-3-11-029535-1,
e-ISBN (EPUB) 978-3-11-038904-3

Group Ring Groups. Orders and Generic Constructions of Units
Eric Jespers, Ángel del Río, 2015
ISBN 978-3-11-037278-6, e-ISBN (PDF) 978-3-11-037294-6,
e-ISBN (EPUB) 978-3-11-038617-2

Group Ring Groups. Structure Theorems of Unit Groups
Eric Jespers, Ángel del Río, 2015
ISBN 978-3-11-041149-2, e-ISBN (PDF) 978-3-11-041150-8,
e-ISBN (EPUB) 978-3-11-041275-8

Joseph Kirtland

Complementation of Normal Subgroups

In Finite Groups

DE GRUYTER

Mathematics Subject Classification 2010
Primary: 20-02, 20E34, 20E07 Secondary: 20D10, 20D25

Author
Prof. Dr. Joseph Kirtland
Marist College
Dept. of Mathematics
3399 North Rd.
Poughkeepsie, NY 12601
USA
joe.kirtland@marist.edu

ISBN 978-3-11-047879-2
e-ISBN (PDF) 978-3-11-048021-4
e-ISBN (EPUB) 978-3-11-047892-1
Set-ISBN 978-3-11-048022-1

Library of Congress Cataloging-in-Publication Data
A CIP catalog record for this book has been applied for at the Library of Congress.

Bibliographic information published by the Deutsche Nationalbibliothek
The Deutsche Nationalbibliothek lists this publication in the Deutsche Nationalbibliografie; detailed bibliographic data are available on the Internet at http://dnb.dnb.de.

Typesetting: Compuscript Ltd., Shannon, Ireland
Printing and binding: CPI books GmbH, Leck
Cover image: Boris SV / gettyimages
♾ Printed on acid-free paper
Printed in Germany

www.degruyter.com

This book is dedicated to Homer Bechtell who ignited my interest in finite group theory and group decompositions. In addition, this text would not have been possible without the loving support of my wife Cindy, son Tim, and daughter Betsy.

Preface

One persistent question in group theory is how to decompose a group into a product of two or more of its subgroups or to determine the properties of a group that is the product of two or more know groups. One basic question from this area is the following: Given a finite group G with normal subgroup N, when does there exist a subgroup H of G, such that $G = NH$ and $N \cap H = \{1\}$. When this occurs, the group G is said to *split* over N by H with H being a *complement* to N in G. This is denoted by $G = [N]H$. This decomposition can provide insights into the structure of the group G, with one advantage being that each element $g \in G$ can be uniquely expressed as $g = nh$ where $n \in N$ and $h \in H$.

When a subgroup H is a complement to a normal subgroup N in group G, then for any element $g \in G$, the subgroup H^g is also a complement to N in G. This motivates the following more general form of the question: Given a normal subgroup N of a group G, when does G split over N, and if G splits over N, when are all of the complements conjugate in G?

Normal subgroup complementation in finite groups is a rich topic that has been investigated by many generations of group theorists including I. Schur, O. Ore, W. Gaschütz, P. Hall, and H. Zassenhaus. This book will document a wide variety of results developed by these and numerous other group theorists. By doing so, the book will also trace the historical development of complementation of normal subgroups in finite groups. By charting the evolution of this topic, the reader will also get a feel for how finite group theory developed over the course of the last 100 years. It is impossible to get a complete understanding of the numerous results mentioned in this book without putting them in the context of the group theoretic developments occurring at the time they were originally proven. Thus, topics such as solvable and nilpotent groups, formations, the Frattini subgroup, system normalizers and many others must be woven into the text. In other words, by tracing the historical development of normal subgroup complementation, the reader will get a front row seat to the development of finite group theory in the twentieth century.

Complementation of normal subgroups in finite groups has its origin in extension theory. In the late nineteenth and early twentieth century, one thrust, initiated by Hölder [63] in 1895, was to classify groups G when given two groups N and H where N is normal in G and $G/N \cong H$. This lead to the development of extension theory and factor systems. Many of the early results concerning normal subgroup complementation were proven using extensions and factor systems.

The main goal of this book is to provide a handy reference for results concerning normal subgroup complementation in finite groups. The goal is not to provide new proofs or new approaches to known results. In fact, most of the proofs included in this text are the original proofs. It should also be mentioned here that not all of the theorems included in this text (e.g. the Odd Order theorem) will be proven. This is

DOI 10.1515/9783110480214-203

due to the fact that some require machinery well beyond the scope of the book or are lengthy and do not add to the discussion.

The numerous lemmas, theorems, and corollaries given in this text will be presented, as their focus allows, in the order in which they were originally proven. When citing results, every effort is made to cite them exactly as they were originally stated. However, since terminology does change over the years, when necessary for clarity, some results are stated using current terminology instead of what was used when originally stated.

While implied by the title of the book and the discussion so far, it should be explicitly stated that for next 140 or so pages, all groups are finite. When a group is mentioned, it will always be assumed to be finite.

The book will be broken down into chapters with central themes. Before any normal subgroup complementation results are presented, Chapter 1 provides a review (or for some an introduction) of the topics from group theory central to the investigation of this topic. Chapter 2 traces the development of the Schur-Zassenhaus theorem. The chapter starts with the famous result by Schur proven in 1902.

Theorem. *If a group G of order nm contains a normal subgroup N of order n, such that N is contained in the center of the G with $(n, m) = 1$, then G is the direct product of N with a subgroup M of order m.*

It then documents the development, from Schur's result to the Odd Order theorem by Feit and Thompson [37], of the Schur-Zassenhaus theorem.

Theorem. *If N is a normal subgroup of a group G and if the order of N and the order of the index of N in G are relatively prime, then there is a complement to N in G and any two such complements are conjugate.*

It ends with a number of generalizations of the Schur-Zassenhaus theorem.

Chapter 3 establishes conditions for when abelian and minimal normal subgroups of a finite group are complemented in the group. Chapter 4, motivated by work by Ore [86] and the seminal paper by Gaschütz [41] presents a series of reduction results concerning normal subgroup complementation. Chapter 5 presents conditions for when normal subgroups in chief series, the derived series, and the lower nilpotent are complemented. Conditions for when normal subgroups with abelian Sylow subgroups are complemented are presented in Chapter 6. This is followed in Chapter 7 where conditions for when $G_{\mathfrak{F}}$, the $\mathfrak{F}$-residual of group G for a saturated formation $\mathfrak{F}$, has a complement in G and when all complements are conjugate. The book ends with Chapter 8, where the structure of groups, with specific classes of subgroups complemented, are examined.

Contents

Notation

(n, m)	Greatest common divisor of n and m
$\mathfrak{A}$	Formation of abelian groups
$\mathcal{A}_n$	Alternating group on n letters
aC	Class of groups where every nontrivial subgroup has a complement
aD	Class of groups with every nontrivial subgroup a direct factor
$Aut(G)$	Automorphism group of a group G
$\mathcal{B}$	Class of groups that split over each normal subgroup not contained in the Frattini subgroup
cC	Class of groups where every nontrivial characteristic subgroup has a complement
cD	Class of groups with every nontrivial characteristic subgroup a direct factor
$C_G(H)$	Centralizer of H in G
$[h, k]$	$h^{-1}k^{-1}hk$ for group elements h and k
$[H, K]$	Subgroup generated by $[h, k]$ for all $h \in H$ and $k \in K$
$x^g = g^{-1}xg$	Conjugation of group element x by element g
$\mathrm{Core}_G(H)$	Largest normal subgroup of G contained in H or $\bigcap_{g \in G} g^{-1}Hg$
C_n	Cyclic group of order n
$D_\infty(G)$	Hypercommutator subgroup of G
$D_{N,\Sigma,\pi}$	$H_\pi \cap \bigcap_{p \in \pi} \mathcal{N}_G({}^pN \cap H_{p'})$, where H_π is the Hall π-subgroup of Σ and $H_{p'}$ is the Hall p'-subgroup of Σ
$\mathcal{F}(G)$	Fitting subgroup of G
$F_i(G)$	$F_i(G) = (F_{i-1}(G))_{\mathfrak{F}}$ for formation $\mathfrak{F}$ with $F_0(G) = G$ and $F_1(G) = G_{\mathfrak{F}}$
$\mathfrak{F}$	Arbitrary formation
$\mathfrak{F}_p$	Formation associated with prime p
Γ	Sylow basis for a group
$\langle S \rangle$	Subgroup of G generated by subset S
$G = [N]H$	Group G splits over N by H ($G = NH$ and $N \cap H = \{1\}$ with $N \trianglelefteq G$)
$[N]_\theta H$	Semidirect product of N with H with $\theta : H \to Aut(N)$
G'	Commutator subgroup of G
$G'(p)$	p-commutator subgroup of group G
$G_{\mathfrak{F}}$	$\mathfrak{F}$-residual of group G in formation $\mathfrak{F}$
$Gl(n, F)$	General linear group of degree n over field F
$Gl(n, p)$	General linear group of degree n over field F with $\|F\| = p$
$H^\#$	Set of non-identity elements in H
H^g	Conjugate of subgroup H by element g in group G or the subgroup $g^{-1}Hg$
H^G	Normal closure of H in group G or the smallest normal subgroup of G containing H

DOI 10.1515/9783110480214-205

H_G	Largest normal subgroup of G contained in H or $\bigcap_{g\in G} g^{-1}Hg$
HK or $H \cdot K$	$\{hk \mid h \in H \text{ and } k \in K\}$
$Inn(G)$	Inner automorphism group of a group G
$[G : H]$	Index of subgroup H in group G
$ker(\theta)$	Kernel of the homomorphism θ
$\mathfrak{N}$	Formation of nilpotent groups
$n\mathfrak{C}$	Formation of solvable nC-groups
nC	Class of groups where every nontrivial normal subgroup has a complement
nD	Class of groups with every nontrivial normal subgroup a direct factor
$\mathcal{N}_G(H)$	Normalizer of subgroup H in G
$\mathcal{N}_G(\Sigma)$	System normalizer for Sylow system Σ of G
${}^pN$	Subgroup of G defined by ${}^pN/N = O_{p'}(G/N)$
$Nr_p\, G$	If ${}^pN$ centralizes no p-chief factor of G of the form N/K
$Nr\, G$	If $Nr_p\, G$ for each prime p dividing $[G : N]$
$Ne\, G$	If ${}^pN$ centralizes no p-chief factor of G below N for each prime p dividing $[G : N]$
$O^\pi(G)$	$\cap\{N \mid N \trianglelefteq G \text{ and } G/N \text{ is a } \pi\text{-group}\}$ or the smallest normal subgroup of G such that $G/O^\pi(G)$ is a π-group.
$O_\pi(G)$	$\langle N \mid N \trianglelefteq G \text{ and } N \text{ is a } \pi\text{-group}\rangle$ or the largest normal subgroup of G which is a π-group
$O_{\pi'\pi}(G)$	$O_\pi(G/O_{\pi'}(G))$ for π a set of primes
$\Phi(G)$	Frattini subgroup of a group G
$\mathfrak{G}_\pi$	Formation of π-groups
π	Set of primes
π'	Set of primes not in π
$\mathbb{P}$	Set of all primes
$R_G(H)$	Intersection of the maximal subgroups in H that are normal in G
Σ	Sylow (Hall) system for a group G
Σ_H	Sylow (Hall) system for subgroup H of group G
Soc(G)	Sockel of G (product of the minimal normal subgroups of G)
S_A	Abelian component of the sockel
S_N	Non-abelian component of the sockel
$\mathfrak{S}$	Formation of solvable groups
$\mathcal{S}_I$	Collection of permutations of set I
$\mathcal{S}_n$	Symmetric group on n letters
$\mathfrak{U}$	Formation of supersolvable groups
$G\, wr\, A$	Wreath product of a nontrivial groups G with respect to A, a subset of $\mathcal{S}_I$
$G \wr H$	Standard wreath product of G with respect to H
$Z(G)$	Center of a group G
$Z_\infty(G)$	Hypercenter of a group G

1 Prerequisites

It is assumed that the reader has a firm understanding of the fundamentals of finite group theory as would be found in a year-long introductory graduate course on the topic. However, some ideas are central to this work and are repeated here. In addition, a few topics required in subsequent chapters, but not typically found in a first course in group theory, are also presented. All groups are finite and the group operation will always be represented multiplicatively. Depending on the circumstances, the identity element of a group will represented by 1 or e.

1.1 Products of subgroups

Definition 1.1. A group G is a **product** of two subgroups A and B if $G = AB$, where $AB = \{ab \mid a \in A \text{ and } b \in B\}$.

When this occurs, each element $g \in G$ can be written, although not necessarily uniquely, as $g = ab$ with $a \in A$ and $b \in B$. In general, given two subgroups A and B of a group G, the product AB need not be a subgroup of G. Clearly, AB is trivially a subgroup of G when $G = AB$. In cases where it may cause confusion, the notation $A \cdot B$ is used to denote AB.

Lemma 1.2. *Let A and B be subgroups of a group G. The product AB is a subgroup of G if and only if $AB = BA$.*

Beweis. Suppose that AB is a subgroup of G, and let $ba \in BA$. Since AB is a subgroup of G, the element $ba = a'b'$ for $a' \in A$ and $b' \in B$. Thus, $BA \subseteq AB$. Now let $ab \in AB$. Thus, $(ab)^{-1} \in AB$ and $(ab)^{-1} = xy$ for $x \in A$ and $y \in B$. Thus, $ab = (xy)^{-1} = y^{-1}x^{-1} \in BA$ and $AB \subseteq BA$.

Now suppose that $AB = BA$. Clearly, AB is nonempty as $ee \in AB$. Let $ab, xy \in AB$, where $a, x \in A$ and $b, y \in B$. Then $(ab)(xy) = a(bx)y = a(x'b')y = (ax')(b'y)$ for $x' \in A$ and $b' \in B$. Thus, AB is closed and a subgroup of G. □

Lemma 1.2 motivates the following definition.

Definition 1.3. Let A and B be subgroups of a group G. The subgroups A and B **permute** if $AB = BA$.

If two subgroups of a group permute, then by Lemma 1.2 their product is a subgroup of the group. Clearly, two subgroups A and B of a group G permute if $G = AB$.

Lemma 1.4. *If A and B are subgroups of a group G with $A \leq \mathcal{N}_G(B)$, then A and B permute.*

DOI 10.1515/9783110480214-001

Beweis. Let $x = ab \in AB$, where $a \in A$ and $b \in B$. It directly follows that $x = ab = aba^{-1}a = b'a$, where $b' \in B$. Thus, $x \in BA$ and $AB \subseteq BA$. A similar calculation shows $BA \subseteq AB$, which implies A and B permute. □

Lemma 1.5. *Let G_1 and G_2 be groups such that $G_1 = AB$, with A and B subgroups of G_1. If $\alpha : G_1 \to G_2$ is a homomorphism, then $\alpha(G_1) = \alpha(A)\alpha(B)$.*

Beweis. Let $g_2 \in \alpha(G_1)$. Then there is an element $g_1 \in G_1$ such that $\alpha(g_1) = g_2$. Thus, $g_1 = ab$, where $a \in A$ and $b \in B$ and

$$g_2 = \alpha(g_1) = \alpha(ab) = \alpha(a)\alpha(b).$$ □

When subgroups A and B of a group G permute, then by Lemma 1.2 the product AB is a subgroup of G, where $|AB| = |A||B|/|A \cap B|$. Permutable subgroups were first studied (as best as the author can tell) by Maillet [80]. Ore [84] also studied them and proved many of their fundamental properties (including the modular identity). For any normal subgroup N and arbitrary subgroup B of a group G, the subgroups N and B always permute by Lemma 1.4 and the product NB is always a subgroup. Now for two famous and extremely useful results.

Lemma 1.6 (The modular identity or Dedekind law). *Let A and B be two subgroups of a group G that permute. If C is a subgroup of G such that $A \leq C$, then $AB \cap C = A(B \cap C)$.*

Beweis. Given that AB is a subgroup of G (Lemma 1.2), $AB \cap C$ is also a subgroup of G. The fact that $A \leq C$ implies that $A(B \cap C) \leq AB \cap C$. Now let $x \in AB \cap C$. Then $x = ab$ for $a \in A$ and $b \in B$. Thus, $a^{-1}x = b \in B \cap C$ and $x = ab \in A(B \cap C)$. Consequently $AB \cap C \leq A(B \cap C)$ and the result follows. □

Lemma 1.7 (The Frattini argument). *If N is a normal subgroup of a group G and $\mathcal{N}_G(P)$ is the normalizer of a Sylow p-subgroup P in N, then $G = N\mathcal{N}_G(P)$.*

Beweis. Let g be an element of G. Since conjugation of the elements of N by g results in an automorphism of N, the subgroup P^g is a Sylow p-subgroup of N. Given that all Sylow p-subgroups of N are conjugate in N, there exists an element $n \in N$ such that $P^g = P^n$. This implies that $P^{gn^{-1}} = P$ and that $gn^{-1} \in \mathcal{N}_G(P)$. Consequently $g = mn$, where $m \in \mathcal{N}_G(P)$. Since $N \trianglelefteq G$, the result that $G = N\mathcal{N}_G(P)$ follows. □

Suppose the group $G = AB$ is the product of subgroups A and B such that $A \cap B = \{1\}$. In this case B is said to be the **complement** of A in G. When this occurs, each element $g \in G$ can be uniquely written as $g = ab$ for $a \in A$ and $b \in B$. If, in addition, A is normal in G, then G is the semidirect product of A with B. This is denoted by $G = [A]_\theta B$

with $\theta : B \to Aut(A)$, where $\theta(b)$ is the automorphism of A induced by conjugation by the element b. Since θ is determined by the group operation, it is usually dropped and $G = [A]B$ is written. If B is also normal in G, then $G = A \times B$ is the direct product of A and B.

It should be noted that a group can have both semidirect and direct product decomposition over the same normal subgroup, as demonstrated in the following example.

Example 1.8. Consider the following group

$$G = \langle a, b \mid a^3 = b^2 = 1, bab = a^2\rangle \times \langle x, y \mid x^3 = y^2 = 1, yxy = y^2\rangle$$

with $N = \langle a, b\rangle$ and $H = \langle x, y\rangle$. It follows directly that $G = N \times H$, but also that $G = [N]D$, where D is the diagonal subgroup with D not normal in G.

If a group G splits over a normal subgroup N by a subgroup H, Bechtell [14] determined the necessary and sufficient conditions when $G = N \times K$ for some subgroup K of G.

In general, given two groups N and H with a homomorphism $\theta : H \to Aut(N)$, the **semidirect product** $G = [N]_\theta H$ can always be formed. For n_1, $n_2 \in N$ and h_1, $h_2 \in H$, the group operation is defined by $n_1h_1 \cdot n_2h_2 = n_1n_2^{\theta(h_1^{-1})}h_1h_2$, where $\theta(h_1^{-1}) \in Aut(N)$. Since the homomorphism θ is arbitrary, for each $h \in H$, $\theta(h)$ can be written as the automorphism of N, where $n^{\theta(h)} = h^{-1}nh = n^h$. Given this, $n_1h_1 \cdot n_2h_2$ can be written as $n_1h_1n_2h_1^{-1}h_1h_2 = n_1n_2^{h_1^{-1}}h_1h_2$.

Definition 1.9. Let N be a normal subgroup of a group G. If there exists a subgroup H of G such that $G = NH$, then G is the **product** of N by H and H is a **supplement** or **partial complement** of N in G. If $N \cap H = \{1\}$, then G **splits** over N by H and is denoted by $G = [N]H$. The subgroup H is a **complement** of N in G and N is **complemented in G.**

Now when a group G does not split over any proper nontrivial normal subgroup, G is said to be **inseparable.** All other groups are **separable.** Inseparable groups have been studied by Bechtell [9, 10, 12], Kirtland [71, 73], and Scarselli [89, 90].

The following lemma (a variation of the Frattini argument (Lemma 1.7))proven by Baer in [1] can now be stated. Its proof is omitted because it is practically identical to the proof of the Frattini argument.

Lemma 1.10 (The Baer argument). *Let $N = [P]Q$ be a normal subgroup of a group G such that P is solvable, $P \le \Phi(G)$, and $(|P|, |Q|) = 1$. Then Q is normal in G and $N = P \times Q$.*

This next lemma contains a host of useful splitting results.

Lemma 1.11. *Let G be a group with subgroups M, N, H, K and L such that M and N are normal in G.*

(i) *If $M \leq N$ and $G = [N]H$, then $G/M = [N/M]HM/M$.*
(ii) *If $M \leq H$ and $G = [N]H$, then $G/M = [NM/M]H/M$.*
(iii) *If $G = [N/M]H/M$ and $N = L \times M$ with $L \lhd G$, then $G = [L]H$.*
(iv) *If $N \leq L$ and $G = [N]H$, then $L = [N](H \cap L)$.*
(v) *If $G = NL$ and $L = [N \cap L]H$, then $G = [N]H$.*
(vi) *If $H \leq L$ and $G = [N]H$, then $L = [N \cap L]H$.*
(vii) *If $G = [N]H$ and $N = M \times L$ with $L \lhd G$, then $G = [M]LH$.*
(viii) *If $G = [N]H$ and $H = [M]K$, then $G = [NM]K$.*
(ix) *If $N \leq M$ such that $G = [N]H$ and M has a complement in G, then M has a complement in G contained in H.*

Beweis. To prove (i), let $gM \in G/M$. Then $g \in G$ and $g = nh$ for $n \in N$ and $h \in H$. Then $gM = nhM = nMhM$ and $G/M = N/M \cdot HM/M$. Suppose that $N/M \cap HM/M \neq \{1_{G/M}\}$. Then there is a nontrivial element $n \in N$ and $h \in H$ such that $n, h \notin M$ and $nM = hM$. This implies that $hn^{-1} \in N$ or that $h \in N$. This is a contradiction as $N \cap H = \{1\}$. Thus, $N/M \cap HM/M = \{1_{G/M}\}$ and $G/M = [N/M]HM/M$.

The proof of (ii) is omitted as it is nearly identical to the proof of (i). To prove (iii), let $g \in G$. Then $gM = nhM$ for $n \in N$ and $h \in H$. Consequently, $(nh)^{-1}g = m$, for $m \in M$, and $g = (nh)m = n(hm) = lm'(hm)$ for $l \in L$ and $m' \in M$. This implies $g = l(m'(hm)) = lh'$, for $h' \in H$, and $G = LH$. Let $x \in L \cap H$. Then $x \in N \cap H = M$. But this means $x \in L \cap M = \{1\}$ and that $x = 1$. Consequently, $G = [L]H$.

For (iv), the modular identity (Lemma 1.6) yields $N(H \cap L) = NH \cap L = G \cap L = L$. Furthermore, $N \cap (H \cap L) = (N \cap H) \cap L = \{1\} \cap L = \{1\}$. As a result, $L = [N](H \cap L)$.

To prove (v), let $g \in G$. Then $g = nl$ for $n \in N$ and $l \in L$. Since $L = [N \cap L]H$, the element $l = n_1h$ for $n_1 \in N \cap L$ and $h \in H$. Then $g = nl = n(n_1h) = (nn_1)h$ and $G = NH$. Let $x \in N \cap H$. Then $x \in H \leq L$ and $x \in L$. This implies that $x \in N \cap H \cap L = \{1\}$. Thus, $N \cap H = \{1\}$ and $G = [N]H$.

For (vi), the modular identity (Lemma 1.6) yields

$$L = G \cap L = HN \cap L = H(N \cap L) = (N \cap L)H.$$

Given that $N \cap L \cap H \leq N \cap H = \{1\}$, it follows that $L = [N \cap L]H$.

For (vii), first note that $L \lhd G$ implies $LH \leq G$. Let $g \in G$. Then $g = nh = (ml)h = m(lh)$ for $n \in N$, $m \in M$, $l \in L$, and $h \in H$, which implies $G = M \cdot LH$. Suppose that exists a nontrivial element $x \in M \cap LH$. This implies $x = m = lh$, where m is nontrivial and l or h is nontrivial. If h is trivial, then $m = l$ and $M \cap L \neq \{1\}$, a contradiction. Thus, h is nontrivial and $l^{-1}m = h$ or $N \cap H \neq \{1\}$, which is a contradiction. As a result, $G = [M]LH$.

For (viii), given that both N and M are normal in G, NM is a normal subgroup of G with $G = (NM)K$. Suppose $NM \cap K \neq \{1\}$. Then there are elements $n \in N$, $m \in M$, and $k \in K$, such that $nm = k$ with $k \neq 1$ and $n \neq 1$ or $m \neq 1$. If $n = 1$, then $m = k$ which contradicts $H = [M]K$. If $m = 1$, then $n = k$, which contradicts $G = [N]H$. Thus, $n = km^{-1}$, which is yet another contradiction. Thus, $NM \cap K = \{1\}$ and $G = [NM]K$.

To show that (ix) holds, let K be a complement to M in G. By the modular identity (Lemma 1.6), it follows that $M \cap NK = NK \cap M = N(K \cap M) = N$. Consider the subgroup $NK \cap H$ of G. First note that

$$M \cap (NK \cap H) = (M \cap NK) \cap H = N \cap H = \{1\}.$$

Given that $G = NH$, the modular identity (Lemma 1.6) implies

$$NK = NK \cap G = NK \cap NH = N(NK \cap H),$$

which results in

$$G = M(NK) = M(N(NK \cap H)) = M(NK \cap H).$$

Thus, $NK \cap H$ is a complement to M in G that is contained in H. □

Two results concerning conjugates of supplements and complements of normal subgroups are now given.

Lemma 1.12. *Let N be a normal subgroup of a group G such that $G = NH$ for a subgroup H of G. Then for any element $g \in G$, the group $G = NH^g$.*

Beweis. By Lemma 1.4, NH^g is a subgroup of G. Let $x \in G$. Then $x^{g^{-1}} \in G$ and $x^{g^{-1}} = nh$ for $n \in N$ and $h \in H$. Thus, $x = n^g h^g = n' h^g$ with $n' \in N$ and $h^g \in H^g$. As a result, $G = NH^g$. □

The following result immediately follows.

Corollary 1.13. *Let N be a normal subgroup of a group G such that $G = [N]H$ for a subgroup H of G. Then for any element $g \in G$, the group $G = [N]H^g$.*

Corollary 1.13 motivates a central question in normal subgroup complementation. If a group $G = [N]H$ for $N, H \leq G$ and $N \trianglelefteq G$, is the collection of subgroups H^g, for $g \in G$, the entire list of complements of N in G? This is trivially the case in the symmetric group $\mathcal{S}_3 = \langle a, b | a^3 = b^2 = 1, bab = a^2 \rangle$, where all of the complements of $\langle a \rangle$ are conjugate as they are the Sylow 2-subgroups of $\mathcal{S}_3$. However, in general, the answer is no. Consider the dihedral group $D_8 = \langle a, b \mid a^4 = b^2 = 1, bab = a^3 \rangle$. In this example, $D_8 = [\langle a \rangle]\langle b \rangle = [\langle a \rangle]\langle ab \rangle$, yet the subgroups $\langle b \rangle$ and $\langle ab \rangle$ are not conjugate in D_8.

This section ends with a simple, yet useful, result concerning subgroup conjugation.

Lemma 1.14. *Let G be a group with subgroups H, L, and N such that $N \le H$, $N \le L$, and $N \trianglelefteq G$. If H/N and L/N are conjugate in G/N, then H and L are conjugate in G.*

Beweis. Let $h \in H$. Then $hN \in H/N$ and there exists an element $gN \in G/N$ such that $(hN)^{gN} \in L/N$ or that $(hN)^{gN} = lN$ for $l \in L$. Therefore, $h^g = ln$ for $n \in N$, which implies that $h^g \in L$. Given that $|H| = |L|$, it follows that H and L are conjugate in G. □

1.2 Series

Definition 1.15. A **subnormal series** for a group G is the chain of subgroups

$$\{1\} = G_0 \trianglelefteq G_1 \trianglelefteq \cdots \trianglelefteq G_k = G,$$

where $G_i \trianglelefteq G_{i+1}$ for $0 \le i \le k-1$. If each factor group is nontrivial and simple, then the series is called a **composition series** for G. If each $G_i \trianglelefteq G$, for $i = 0, \ldots, k-1$, the series is called a **normal series.** A **principal series** or **chief series** is a normal series where for each $i = 0, \ldots, k-1$, the subgroup G_i is maximal among normal subgroups of G properly contained in G_{i+1}. In this case G_{i+1}/G_i is called **chief factor** of G. If each G_i, for $i = 0, \ldots, k-1$, is characteristic in G and G_i is the maximal characteristic subgroup of G contained in G_{i+1}, then the series is called a **characteristic series** for G.

Definition 1.16. A subgroup of a group is **subnormal** in that group if it is contained in a subnormal series for that group.

Every finite group has a composition series and any two composition series for a group G are equivalent (the Jordan-Hölder theorem). A group G is **solvable** if it has a subnormal series whose factors are all abelian. This is equivalent to G having a chief series with each chief factor an elementary abelian p-group. If each chief factor of a solvable group is of prime order, then the group G is **supersolvable.**

Let $\mathbb{P}$ be the set of all primes, and let π be a subset of $\mathbb{P}$. Then π' denotes the set of primes in $\mathbb{P}$, but not in π. A positive integer n is called a **π-number** if each of the primes that divide n are in π. A group G is a **π-group** if its order is a π-number. It follows directly that subgroups and homomorphic images of π-groups are also π-groups. A group G is **π-separable** if each chief factor in a chief series for G is either a π- or π'-group and is **π-solvable** if in addition each π-chief factor is abelian.

Before returning to the discussion on series, the introduction of π-groups leads to the following definition.

Definition 1.17. Let π be a set of primes, and let G be a group. The subgroup

$$O^{\pi}(G) = \bigcap_i \{N_i \mid N_i \trianglelefteq G \text{ and } G/N_i \text{ is a } \pi\text{-group}\}.$$

Equivalently, $O^{\pi}(G)$ is the smallest normal subgroup of G such that $G/O^{\pi}(G)$ is a π-group. It is obvious that $O^{\pi}(G)$ is characteristic in G. Now let N and H be normal π-subgroups of a group G. Then NH is a normal π-subgroup of G as $|NH| = |N||H|/|N \cap H|$. This motivates the following definition.

Definition 1.18. Let G be a group, and let π be a set of primes. Then $O_{\pi}(G)$ is the largest normal π-subgroup of G.

Again, it is obvious that $O_{\pi}(G)$ is characteristic in G.

Definition 1.19. Let G be a group, and let π be a set of primes. Then

$$O_{\pi'\pi}(G) = O_{\pi}(G/O_{\pi'}(G)).$$

Note when $\pi = p$, where p is a prime, that $O_{p'p}(G)$ is the largest normal p-nilpotent subgroup of G (see Definition 1.48 in Section 1.4 for the definition of p-nilpotent). If G is a solvable group, then $O_{p'p}(G)$ is also the intersection of the centralizers of the p-chief factors of G.

Definition 1.20. The **upper central series** for a group G is the chain of subgroups

$$\{1\} = Z_0(G) \leq Z_1(G) \leq \cdots \leq Z_k(G) \leq \cdots,$$

where $Z_1(G) = Z(G)$, and for $i \geq 2$, $Z_i(G)/Z_{i-1}(G) = Z(G/Z_{i-1})$. The subgroup $Z_{\infty}(G) = \cup_i Z_i(G)$ is the **hypercenter** of G.

If $Z_{\infty}(G) = G$ for a group G, then G is **nilpotent** and the length of the upper central series is the **class** of the nilpotent group.

Definition 1.21. The **lower central series** for a group G is the chain of subgroups

$$G = D_0 \geq D_1 \geq \cdots \geq D_k \geq \cdots,$$

where for $i \geq 1$, $D_i = [D_{i-1}, G]$. The **hypercommutator subgroup** $D_{\infty}(G) = \cap_i D_i(G)$.

Note that $D_{\infty}(G)$ is the (unique) smallest normal subgroup of G such that $G/D_{\infty}(G)$ is nilpotent. If a group G is nilpotent, then $D_{\infty}(G) = \{1\}$.

Definition 1.22. The **lower nilpotent series** for a group G is the chain of subgroups

$$G = L_0 \geq L_1 \geq \cdots \geq L_k \geq \cdots ,$$

where for $i \geq 1$, $L_i = D_\infty(L_{i-1})$.

A group G is solvable if and only if $L_n = \{1\}$ for some $n \geq 1$.

Definition 1.23. The **upper nilpotent series** for a group G is the chain of subgroups

$$\{1\} = \mathcal{F}_0(G) \leq \mathcal{F}_1(G) \leq \cdots \mathcal{F}_k(G) \leq \cdots ,$$

where for $i \geq 0$, $\mathcal{F}_{i+1}(G)/\mathcal{F}_i(G) = \mathcal{F}(G/\mathcal{F}_i(G))$, with $\mathcal{F}(G/\mathcal{F}_i(G))$ being the Fitting subgroup of $G/\mathcal{F}_i(G)$.

See Section 1.6 for information concerning the Fitting subgroup. Also note that a group G is solvable if and only if $\mathcal{F}_n(G) = G$ for some $n \geq 1$.

Definition 1.24. The **derived series** for a group G is the series

$$G = G^0 \geq G^1 \geq G^2 \geq \cdots \geq G^k \geq \cdots ,$$

where $G^1 = G'$ and for $i \geq 2$, $G^i = (G^{i-1})'$.

A group G is solvable if and only if $G^n = \{1\}$ for some $n \geq 1$. If $G = G'$, then G is a **perfect** group.

1.3 Solvable groups, chief factors, sylow systems, and system normalizers

Recall that a group G is **solvable** if it has a subnormal series

$$\{1\} = G_0 \trianglelefteq G_1 \trianglelefteq \cdots \trianglelefteq G_k = G$$

whose factors are all abelian. This is equivalent to G having a chief series

$$\{1\} = H_0 \triangleleft H_1 \triangleleft \cdots \triangleleft H_r = G$$

with each chief factor H_i/H_{i-1} an elementary abelian p-group or to G having a derived series with $G^n = \{1\}$ for some $n \geq 1$.

Lemma 1.25. *Let G be a group.*
(i) *If G is solvable, then any subgroup and homomorphic image of G is also solvable.*
(ii) *If N and G/N are solvable for $N \trianglelefteq G$, then G is solvable.*

Beweis. To prove (i), first note that since G is solvable, there is a subnormal series $\{1\} = G_0 \trianglelefteq G_1 \trianglelefteq \cdots \trianglelefteq G_k = G$ such that G_{i+1}/G_i is abelian for $0 \leq i \leq k-1$. Let H be a subgroup of G and consider the subgroups $H_i = G_i \cap H$. Each H_i is normal in H_{i+1}, with $H_{i+1} \cap G_i = H_i$. Given that

$$H_{i+1}/H_i = H_{i+1}/(H_{i+1} \cap G_i) \cong H_{i+1}G_i/G_i \leq G_{i+1}/G_i,$$

each factor in the subnormal series $\{1\} = H_0 \trianglelefteq H_1 \trianglelefteq \cdots \trianglelefteq H_k = H$ is abelian and H is solvable.

Now let N be a normal subgroup of G and consider the subgroups $N_i = NG_i$. For $0 \leq i \leq k-1$, each subgroup N_i is normal in N_{i+1}, with $N_{i+1} = N_iG_{i+1}$. It then follows that

$$N_{i+1}/N_i = N_iG_{i+1}/NG_i = (NG_i)G_{i+1}/NG_i$$

and that

$$N_{i+1}/N_i \cong G_{i+1}/(NG_i \cap G_{i+1}) \cong (G_{i+1}/G_i)/((NG_i \cap G_{i+1})/G_i).$$

This implies

$$\{1_{G/N}\} = N_0/N \trianglelefteq N_1/N \trianglelefteq \cdots \trianglelefteq N_k/N = G/N$$

is a subnormal series for G/N with abelian factor groups or that G/N is solvable.

For (ii), let $\alpha : G \to G/N$ be the natural homomorphism from G onto G/N. Since G/N is solvable, there is a subnormal series $\{1_{G/N}\} = G_0/N \trianglelefteq G_1/N \trianglelefteq \cdots \trianglelefteq G_k/N = G/N$ for G/N with abelian factors. In addition, there is a subnormal series $\{1\} = N_0 \trianglelefteq N_1 \trianglelefteq \cdots \trianglelefteq N_l = N$ for N with abelian factors. Thus,

$$\{1\} = N_0 \trianglelefteq N_1 \trianglelefteq \cdots \trianglelefteq N_l \trianglelefteq \alpha^{-1}(G_1) \trianglelefteq \cdots \trianglelefteq \alpha^{-1}(G_k) = G$$

is a subnormal series for G with abelian factor groups and G is solvable. □

An integral part of any study of solvable groups are Hall subgroups, introduced by Hall [51].

Definition 1.26. Let π be a set of primes. A subgroup H of a group G is called a **Hall π-subgroup** if $|H|$ is a π-number and $[G : H]$ is a π'-number. A subgroup is called a **Hall subgroup** if it is a Hall π-subgroup for some $\pi \subseteq \mathbb{P}$.

Lemma 1.27. *Let H be a Hall π-subgroup of a group G, and let $N \trianglelefteq G$.*
(i) *H^g is a Hall π-subgroup for all $g \in G$,*
(ii) *HN/N is a Hall π-subgroup of G/N, and*
(iii) *$H \cap N$ is a Hall π-subgroup of N.*

Beweis. Item (i) follows from the fact that $|H| = |H^g|$. For (ii), first note that HN/N is π-subgroup of G/N as $|HN/N| = |H/N \cap H|$. Since $[G/N : HN/N] = [G : HN]$, $[G/N : HN/N]$ is a π'-number and HN/N is a Hall π-subgroup of G/N. Item (iii) follows from the fact that $[HN : H] = [N : N \cap H]$. □

Theorem 1.28. [Theorems 2.1 and 2.2 in Hall [51]] *Let G be a solvable group and π a set of primes. Then*
(i) *Hall π-subgroups of G exist and*
(ii) *any two Hall π-subgroups are conjugate.*

Beweis. Assume that G is nontrivial and that its order is divisible by two or more primes. Proceed by induction on the order of G. Let N be a minimal normal subgroup of G. Then N is an elementary abelian p-group.

To prove (i), first note that by induction G/N has a Hall π-subgroup H/N. If $p \in \pi$, then H is a Hall π-subgroup for G. If $p \notin \pi$, then, by Theorem 2.12, $H = [N]K$, where K is a π-subgroup. Given that $[G : K] = [G : H][H : K] = [G : H]|N|$, the subgroup K is a Hall π-subgroup of G.

For (ii), let H and K be Hall π-subgroups of G. Then HN/N and KN/N are Hall π-subgroups of G/N. By induction, there exists an element $gN \in G/N$ such that $(HN/N)^{gN} = KN/N$ or $H^gN = KN$. If $p \in \pi$, then $K = KN = H^gN = H^g$. If $p \notin \pi$, then $KN = [N]K = [N]H^g$. By Theorem 2.13, the subgroups K and H^g are conjugate in KN. Thus, K and H are conjugate in G. □

The existence of Hall π-subgroups in solvable groups leads to Sylow systems which were introduced by Hall [54].

Definition 1.29. Let G be a solvable group. A **Sylow** or **Hall system** of G is the set Σ of Hall subgroups of G that satisfy the following two properties:
(i) For each $\pi \subseteq \mathbb{P}$, Σ contains exactly one Hall π-subgroup of G.
(ii) If $H, K \in \Sigma$, then $HK = KH$.

Hall [54] also proved that a group G is solvable if and only if G has a Hall system. Given a Hall system Σ for a solvable group G and H a Hall π-subgroup in Σ, there exists, by definition, a Hall π'-subgroup $K \in \Sigma$ such that $HK = KH$. For any solvable subgroup L of a group G, a Sylow system for L is denoted by Σ_L.

Lemma 1.30. *Let N be a solvable normal subgroup of a group G.*
(i) *If $\Sigma_N = \{S_1, \dots, S_r\}$ is a Sylow system for N, then for $g \in G$, $\Sigma_N^g = \{S_1^g, \dots, S_r^g\}$ is also a Sylow systems for N.*
(ii) *Any two Sylow systems for N are conjugate in N.*

Beweis. For (i), first note that $S_i^g \le N$ with $|S_j^g| = |S_i|$ for $1 \le i \le r$. Thus, Σ_N^g contains exactly one Hall π-subgroup of N for each $\pi \in \mathbb{P}$. Since

$$S_i^g S_j^g = g^{-1}S_i g g^{-1}S_j g = g^{-1}S_i S_j g = g^{-1}S_j S_i g = g^{-1}S_j g g^{-1}S_i g = S_j^g S_i^g$$

for any $S_i, S_j \in \Sigma_N$, it follows that Σ_N^g is a Sylow system for N.

The proof of (ii), while fairly straightforward, is omitted (see [34]). □

It should also be mentioned that if Σ_H is a Sylow system for a subgroup H of a solvable group G, there is a Sylow system $\Sigma = \{S_1, \dots, S_r\}$ for G such that

$$\Sigma_H = H \cap \Sigma = \{H \cap S_1, \dots, H \cap S_r\}.$$

A Sylow system Σ for a solvable group G is said to **reduce into** H, where H is a subgroup of G when $H \cap \Sigma$ is a Sylow system for H. Note that given an arbitrary Sylow system for G, it does not necessarily reduce into H. However, as noted above, given a Sylow system for H, there always exists a Sylow system for G that reduces into H.

Hall [55] introduced the concept of a system normalizer.

Definition 1.31. Let N be a solvable normal subgroup of a group G, and let $\Sigma_N = \{S_1, \dots, S_r\}$ be a Sylow system for N.
(i) The **relative system normalizer of N (relative to G)** is $\mathcal{N}_G(\Sigma_N) = \{g \in G \mid g^{-1}S_i g = S_i \text{ for } i = 1, \dots, r\}$.
(ii) The **absolute system normalizer of N (relative to G)** is $\mathcal{N}_N(\Sigma_N) = N \cap \mathcal{N}_G(\Sigma_N)$.
(iii) If $N = G$, then $\Sigma = \Sigma_N$ is a Sylow system for G and $\mathcal{N}_G(\Sigma)$ is the **system normalizer** of Σ in G.

The system normalizers of a solvable group G are nilpotent, form a characteristic conjugacy class of subgroups, and the number of Hall systems is the index of a system normalizer in G. In addition, the intersection of all of the system normalizers of a solvable group G is the hypercenter. While some basic results concerning system normalizers are presented in this section, an in-depth discussion of this topic can be found in Doerk and Hawkes [34].

Lemma 1.32. *Let Σ be a Sylow system for a solvable group G, and let P be a Sylow p-subgroup in Σ and S a p'-subgroup in Σ. Then $P \cap \mathcal{N}_G(S)$ is a Sylow p-subgroup of $\mathcal{N}_G(S)$ and $\mathcal{N}_G(\Sigma)$.*

Beweis. First note that $G = PS = P\mathcal{N}_G(S)$. By the modular identity (Lemma 1.6)

$$(P \cap \mathcal{N}_G(S))S = PS \cap \mathcal{N}_G(S) = \mathcal{N}_G(S),$$

which implies $|P \cap \mathcal{N}_G(S)| = [\mathcal{N}_G(S) : S]$ and that $P \cap \mathcal{N}_G(S)$ is a Sylow p-subgroup of $\mathcal{N}_G(S)$. It then follows directly that $P \cap \mathcal{N}_G(S)$ is a Sylow p-subgroup of $\mathcal{N}_G(\Sigma)$. □

This next result is motivated by the Frattini argument (Lemma 1.7).

Lemma 1.33. *For any solvable normal subgroup N of a group G, $G = N\mathcal{N}_G(\Sigma_N)$.*

Beweis. Let $\Sigma_N = \{S_1, \ldots, S_r\}$ be a Sylow system for N and $g \in G$. Then by Lemma 1.30, Σ_N^g is a Sylow system of N and there is an $n \in N$ such that $\Sigma_N^g = \Sigma_N^n$. Thus, $\Sigma_N^{gn^{-1}} = \Sigma_N$ and $gn^{-1} = m \in \mathcal{N}_G(\Sigma_N)$. Thus, $g = mn$ and $G = N\mathcal{N}_G(\Sigma_N)$. □

Before continuing, one last property concerning solvable groups must be mentioned. Given a solvable group G, it has a Sylow system Σ. For each prime dividing the order of G, there is a Sylow p-subgroup S_p in Σ. Furthermore, $S_pS_q = S_qS_p$ for any two Sylow subgroups in Σ. Thus, if $\{S_{p_1}, \ldots, S_{p_t}\}$ is the set of Sylow subgroups in Σ, then $G = S_{p_1} \cdots S_{p_t}$. This motivates the following definition and theorem.

Definition 1.34. A set Γ consisting of exactly one Sylow p-subgroup for each prime p dividing the order of G, where subgroups in Γ are pairwise permutable, is a **Sylow basis** for G.

Theorem 1.35. *Let G be a solvable group with $|G| = p_1^{r_1} \cdots p_t^{r_t}$, then G has at least one Sylow basis $\Gamma = \{S_{p_1}, \ldots, S_{p_t}\}$, where $G = S_{p_1} \cdots S_{p_t}$.*

Lemma 1.36. *Let G be a solvable group with Sylow basis $\Gamma = \{S_{p_1}, \ldots, S_{p_t}\}$, and let N be a normal subgroup of G. Then the collection of subgroups $N \cap S_{p_i}$, for $1 \le i \le t$, where p_i divides the order of N, is a Sylow basis for N.*

Beweis. Clearly, for each p, where p divides the order of N, $N_p = N \cap S_p$ is a Sylow p-subgroup of N. Consider the Sylow subgroups N_p and N_q of N. Clearly, by their definition, it follows that $N_pN_q \le N \cap S_pS_q$. However, $N \cap S_pS_q$ is a subgroup of N of order $p^\alpha q^\beta$ for $\alpha, \beta \ge 0$. This means that the order of $N \cap S_pS_q$ divides $|N_p|\,|N_q|$. Consequently, using Lemma 1.2, it follows that

$$N_pN_q = N \cap S_pS_q = N_qN_p$$

and that collection of subgroups $N \cap S_{p_i}$, for $1 \le i \le t$, where p_i divides the order of N, is a Sylow basis for N. □

Lemma 1.37. *Let G be a solvable group with Sylow basis $\Gamma = \{S_{p_1}, \dots, S_{p_t}\}$, and let N be a normal subgroup of G. If an element $g = s_1 \cdots s_t \in N$, where $s_j \in S_{p_j}$ for $1 \le j \le t$, then each $s_j \in N$ for $1 \le j \le t$.*

Beweis. First note that by Lemma 1.36, $N = \left(N \cap S_{p_1}\right) \cdots \left(N \cap S_{p_t}\right)$. Proceed by induction on t. Thus, $g = s_1 \cdots s_t = x_1 \cdots x_t$, where $x_j \in N \cap S_{p_j}$ for $1 \le j \le t$. This implies

$$x_1^{-1} s_1 = (x_2 \cdots x_t)(s_2 \cdots s_t)^{-1}.$$

Given that $S_1 \cap S_2 \cdots S_t = \{1\}$, it follows that $x_1^{-1} s_1 = 1$ or that $x_1 = s_1$, forcing $s_1 \in N$. By induction, $s_j \in N$ for $2 \le j \le n$. □

Definition 1.38. Let H/K be a chief factor for a group G, and let L be a subgroup of G.
(i) L **covers** H/K if $H \le KL$ (which is equivalent to $KL \cap H = H$ or $HL = KL$). L **avoids** H/K if $L \cap H \le K$ (which is equivalent to $KL \cap H = K$ or $H \cap L = K \cap L$).
(ii) The chief factor H/K is a **central chief factor** if $H/K \le Z(G/K)$. If H/K is not central, then it is an **eccentric chief factor**.

Theorem 1.39. *Let G be a solvable group. Every system normalizer of G covers every central chief factor and avoids every eccentric chief factor.*

Beweis. Let Σ be a Sylow system for G, and let H/K be a chief factor of G with $|H/K| = p^\alpha$ for some prime p. Let P be a Sylow p-subgroup of G such that $P \in \Sigma$, and let S be the Hall p'-subgroup in Σ.

Case 1: H/K is central. In this case, $HS/K = KS/K \times H/K$, where $[HS : KS] = p^\alpha$. Note that S is a Hall p'-subgroup of KS. Let $h \in HS$. Since $KS \lhd HS$, S^h is a Hall p'-subgroup of SK. By Theorem 1.28, there is an element $k \in SK$ such that $S^h = S^k$ or that $S^{hk^{-1}} = S$. This implies that $hk^{-1} \in \mathcal{N}_{SH}(S)$ or that $hk^{-1} = l$, where $l \in \mathcal{N}_{SH}(S)$. Consequently, $h = lk$ and $HS = KS\mathcal{N}_{HS}(S)$.

Since S and $\mathcal{N}_{HS}(S)$ are both subgroups of $\mathcal{N}_G(S)$, it follows that $HS \le K\mathcal{N}_G(S)$. Thus, H/K, which is a normal p-subgroup of G/K, satisfies $H/K \le K\mathcal{N}_G(S)/K$. This implies $H/K \trianglelefteq K\mathcal{N}_G(S)/K$ and that H/K is contained in every Sylow p-subgroup of $K\mathcal{N}_G(S)/K$. By Lemma 1.32, $P^* = P \cap \mathcal{N}_G(S) \le \mathcal{N}_G(\Sigma)$ and is a Sylow p-subgroup of $\mathcal{N}_G(S)$. This means P^*K/K is a Sylow p-subgroup of $K\mathcal{N}_G(S)/K$. Given that $H/K \le P^*K/K \le \mathcal{N}_G(\Sigma)K/K$, it follows that $H \le \mathcal{N}_G(\Sigma)K$ or that $\mathcal{N}_G(\Sigma)$ covers H/K.

Case 2: H/K is eccentric. Here H/K is a minimal normal subgroup of G/K with SK/K (by Lemma 1.27) a Hall p'-subgroup of G/K. If $H/K \cap \mathcal{N}_{G/K}(SK/K) \neq \{1\}$, then $H/K \le Z(G/K)$ (see I.5.5 in [34]) and is central. This contradiction implies that $H/K \cap \mathcal{N}_{G/K}(SK/K) = \{1\}$.

Given that H/K is a normal p-subgroup of G/K, $H/K \cap \mathcal{N}_G(\Sigma)K/K \leq P_1K/K$, where P_1K/K is a Sylow p-subgroup of $\mathcal{N}_G(\Sigma)K/K$. Given that $P_1 \leq \mathcal{N}_G(\Sigma)$, it follows that $P_1 \leq \mathcal{N}_G(S)$, which implies $P_1K/K \leq \mathcal{N}_{G/K}(SK/K)$. Since $H/K \cap \mathcal{N}_{G/K}(SK/K) = \{1\}$, this means $H/K \cap \mathcal{N}_G(\Sigma)K/K = \{1\}$ or that $H \cap \mathcal{N}_G(\Sigma)K = K$. Thus, $\mathcal{N}_G(\Sigma)$ avoids H/K. □

This next result is one of the direct consequences of Theorem 1.39.

Theorem 1.40. *Let $\mathcal{N}_G(\Sigma)$ be a system normalizer of a solvable group G. Then $Core_G(\mathcal{N}_G(\Sigma)) = Z_\infty(G)$.*

Beweis. Let $C = \mathrm{Core}_G(\mathcal{N}_G(\Sigma))$. Since all of the chief factors for G contained in $Z_\infty(G)$ are central, Theorem 1.39 implies they are covered by $\mathcal{N}_G(\Sigma)$. Thus, $Z_\infty(G) \leq \mathcal{N}_G(\Sigma)$, which implies that $Z_\infty(G) \leq C$.

Now let

$$\{1\} = C_0 \lhd C_1 \lhd \cdots \lhd C_t = C$$

be part of a chief series for G containing C. Given that $\mathcal{N}_G(\Sigma)$ covers each chief factor C_i/C_{i-1}, for $1 \leq i \leq t$, it follows from Theorem 1.39 that each of these chief factors in central. This means $[C_i, G] \leq C_{i-1}$, which results in $C \leq Z_\infty(G)$. Consequently, it follows that $\mathrm{Core}_G(\mathcal{N}_G(\Sigma)) = Z_\infty(G)$. □

1.4 Supersolvable groups

Definition 1.41. A group G is **supersolvable** if it has a chief series

$$\{1\} = H_0 \lhd H_1 \lhd \cdots \lhd H_r = G$$

such that each chief factor H_i/H_{i-1} is cyclic of prime order.

Theorem 1.42. *Let G be a supersolvable group.*
(i) *Every subgroup and homomorphic image of G is supersolvable.*
(ii) *Each maximal subgroup of G has prime index in G.*

Beweis. The proof of (i) is omitted as it is almost identical to the proof of (i) in Lemma 1.25. For (ii), let M be a maximal subgroup on G. If M is normal in G, then it follows directly that it has prime index in G. Suppose M is not normal in G, and let $N = \mathrm{Core}_G(M)$.

By (i), G/N is supersolvable with $\mathrm{Core}_{G/N}(M/N) = \{1_{G/N}\}$. Thus, there exists a minimal normal subgroup H/N of G/N such that $H/N \cap M/N = \{1_{G/N}\}$ with

$|H/N| = p$, for p a prime. Given that $G/N = H/N \cdot M/N$, it follows that $G/N = [H/N]M/N$. Consequently, $[G : M] = [G/N : M/N] = |H/N| = p$. □

As demonstrated by the symmetric group $\mathcal{S}_4$, given a group G and normal subgroup N of G such that N and G/N are supersolvable does not necessarily imply that G is supersolvable.

Theorem 1.43. *Let p be the largest prime dividing the order of a supersolvable group G. Then the Sylow p-subgroup P of G is normal in G.*

Beweis. Suppose that P is not normal in G. Then $\mathcal{N}_G(P) \leq M$ for some maximal subgroup M of G. Note that this implies $\mathcal{N}_G(P) = \mathcal{N}_M(P)$. It directly follows that

$$[G : \mathcal{N}_G(P)] = [G : M][M : \mathcal{N}_G(P)] = [G : M][M : \mathcal{N}_M(P)].$$

By Sylow's theorem, this implies $[G : M] = kp + 1$ for some $k \geq 1$. By Theorem 1.42, it follows that $[G : M] = q$ for some prime q dividing the order of G. This is a contradiction, as $p > q$, which implies that P is normal in G. □

Theorem 1.44. *For a supersolvable group G, the subgroup G′ is nilpotent.*

Beweis. Let

$$\{1\} = H_0 \lhd H_1 \lhd \cdots \lhd H_r = G$$

be a chief series for G. Let $N_i = H_i \cap G'$ for $1 \leq i \leq r$. Consider the homomorphism $\alpha : G/N_i \to Aut(N_{i+1}/N_i)$, where $\alpha(gN_i)$, for $g \in G$, is the automorphism of N_{i+1}/N_1 determined by conjugation by gN_i. The kernel of α is $K/N_i = C_{G/N_i}(N_{i+1}/N_i)$. Thus, G/K is isomorphic to a subgroup of $Aut(N_{i+1}/N_i)$. However, given that N_{i+1}/N_i is cyclic, $Aut(N_{i+1}/N_i)$ is abelian, which means $G' \leq K$. Consequently, $N_{i+1}/N_1 \leq Z(G'/N_i)$, which implies that G' is nilpotent. □

1.5 Nilpotent groups

Recall that a group G is **nilpotent** if it has an upper central series

$$\{1\} = Z_0(G) \leq Z_1(G) \leq \cdots \leq Z_r(G) = G,$$

where $Z_r(G) = G$ for some positive integer r. Given that the center of a p-group is nontrivial and that the homomorphic image of a p-group is a p-group, any p-group is nilpotent.

Lemma 1.45. *Let G be a nilpotent group.*
(i) *Every subgroup and homomorphic image of G is nilpotent.*
(ii) *If N is a nontrivial normal subgroup of G, then $N \cap Z(G) \neq \{1\}$.*
(iii) *If H is a proper subgroup of G, then $H < \mathcal{N}_G(H)$.*
(iv) *All maximal subgroups of G are normal in G.*

Beweis. Let

$$\{1\} = Z_0(G) \leq Z_1(G) \leq \cdots \leq Z_r(G) = G$$

be the upper central series for G.

For (i), let H be a subgroup of G. Since $H \cap Z_i(G) \leq Z_i(H)$, for $0 \leq i \leq k$, it follows that H is nilpotent. Now let N be a normal subgroup of G. Given that $Z_i(G)N/N \leq Z_i(G/N)$, G/N is nilpotent.

To prove (ii), suppose that $N \cap Z(G) = \{1\}$. Then for some i, where $2 \leq i \leq r$, $N \cap Z_i(G) \neq \{1\}$ with $N \cap Z_{i-1}(G) = \{1\}$. Let $n \in N \cap Z_i(G)$ and $g \in G$. Then $nZ_{i-1}(G)gZ_{i-1}(G) = gZ_{i-1}(G)nZ_{i-1}(G)$ or $n^{-1}g^{-1}ng \in Z_{i-1}(G)$. Given that N is normal in G, this implies $N \cap Z_{i-1}(G) \neq \{1\}$, a contradiction. Thus, $N \cap Z(G) \neq \{1\}$ when $N \neq \{1\}$.

For (iii), there exists an i, with $0 \leq i \leq r-1$, such that $Z_i(G) \leq H$ but $Z_{i+1}(G) \nleq H$. This implies $[Z_{i+1}(G), H] \leq [Z_{i+1}(G), G] \leq Z_i(G)$. Thus, there exists an element $x \in Z_{i+1}(G)$ such that $x \notin H$ and $[x, h] \in H$ for all $h \in H$. Consequently, $H^x = H$ and $H < \mathcal{N}_G(H)$.

Item (iv) follows directly from (iii). □

Note that given a group G with $N \trianglelefteq G$, the condition N and G/N nilpotent does not necessarily imply that the group G is nilpotent. The alternating group $\mathcal{A}_4$ is a counterexample.

Theorem 1.46. *A group G is nilpotent if and only if G is the direct product of its Sylow p-subgroups (i.e. each Sylow p-subgroup is normal in the group).*

Beweis. First assume G is nilpotent, and let P be a Sylow p-subgroup of G. Suppose that $\mathcal{N}_G(P) \neq G$. Then $\mathcal{N}_G(P) \leq M$ for some maximal subgroup M of G. Thus, P is a Sylow p-subgroup of M, which by Lemma 1.45 is normal in G. Thus, by the Frattini argument (Lemma 1.7), $G = M\mathcal{N}_G(P)$. This is a contradiction. Thus, $\mathcal{N}_G(P) = G$ and P is normal in G, which implies G is the direct product of its Sylow p-subgroups. The converse follows from the fact that the center of a p-group is nontrivial. □

The next result follows directly from Theorem 1.46.

Theorem 1.47. *If all of the maximal subgroups of a group G are normal, then G is nilpotent.*

Beweis. Let P be a Sylow p-subgroup of G and consider $\mathcal{N}_G(P)$. If $\mathcal{N}_G(P) \neq G$, then $\mathcal{N}_G(P) \leq M$ for some maximal subgroup M of G. Thus, by the Frattini argument (Lemma 1.7), $G = N\mathcal{N}_G(P)$. This contradiction implies that $\mathcal{N}_G(P) = G$, and using Theorem 1.46, that G is nilpotent. □

While Theorem 1.46 gives a classification of nilpotent groups, it also motivates the following definition.

Definition 1.48. Let p be a prime. A group G is ***p*-nilpotent** if for for some Sylow p-subgroup P of G and normal subgroup N of G, $G = [N]P$.

An equivalent condition for a group G to be p-nilpotent is that $O_{p'p}(G) = G$. Subgroups and homomorphic images of p-nilpotent groups are also p-nilpotent.

1.6 The Frattini and Fitting subgroups

Definition 1.49. An element g of a group G is a **nongenerator** of G if for all subsets S of G such that $G = \langle g, S\rangle$, then $G = \langle S\rangle$.

Definition 1.50. The **Frattini subgroup** of a group G, denoted by $\Phi(G)$, is the set of nongenerators of G.

The Frattini subgroup was introduced and studied by Giovanni Frattini [38, 39]. If the group $G = \{1\}$, then $\Phi(G) = \{1\}$. Otherwise, $\Phi(G)$ is a proper subgroup of G. The definition implies, without justification, the fact that $\Phi(G)$ is a subgroup of the group. Clearly, the identity element of the group G is in $\Phi(G)$. Now let $g, h \in \Phi(G)$ and suppose that $G = \langle gh, S\rangle$ for some subset S of G. Then it follows that

$$\langle gh, S\rangle \leq \langle g, h, S\rangle \leq \langle h, S\rangle = \langle S\rangle = G$$

and that $gh \in \Phi(G)$. Thus, $\Phi(G)$ is a subgroup of G.

Lemma 1.51. *Let G be a group.*
(i) *$\Phi(G)$ is the intersection of the maximal subgroups of a group G.*
(ii) *$\Phi(G)$ is nilpotent.*
(iii) *If a prime p divides $|\Phi(G)|$, then p divides $|G/\Phi(G)|$.*

Beweis. If $G = \{1\}$, then the result follows. Assume that G is nontrivial. To prove (i), let $x \in \Phi(G)$, and suppose there exists a maximal subgroup M of G such that $x \notin M$. Then it follows that $M < \langle x, M\rangle = G$. But since x is a nongenerator, this implies that $\langle x, M\rangle = M = G$, a contradiction. Thus, $x \in M$ and $\Phi(G)$ is contained in the intersection of the maximal subgroups.

Now let x be an element of the intersection of the maximal subgroups of G. Suppose there is a subset S of G such that $G = \langle x, S\rangle$ and $G \neq \langle S\rangle$. Let M be a subgroup of G and maximal with respect to $\langle S\rangle \leq M$ and $x \notin M$. If M is not a maximal subgroup of G, then $M < K$, where K is maximal in G. But $x \in K$, which implies that $\langle x, S\rangle \leq K$ or $K = G$, a contradiction. This implies M is a maximal subgroup of G with $x \notin M$, another contradiction. As a result, $x \in \Phi(G)$ and the intersection of the maximal subgroups of G is contained in $\Phi(G)$.

For (ii), first note that by (i) that $\Phi(G)$ is a characteristic subgroup of G. If $\Phi(G) = \{1\}$, the result follows. Assume that $\Phi(G)$ is nontrivial and let P be a Sylow p-subgroup of $\Phi(G)$. By the Frattini argument (Lemma 1.7), it follows that $G = \Phi(G)\mathcal{N}_G(P)$. If $\mathcal{N}_G(P)$ is a proper subgroup of G, then $G = \langle\Phi(G), \mathcal{N}_G(P)\rangle = \langle\mathcal{N}_G(P)\rangle = \mathcal{N}_G(P)$, which is a contradiction. Thus, $\mathcal{N}_G(P) = G$ and $P \lhd G$. This implies that every Sylow subgroup of $\Phi(G)$ is normal in $\Phi(G)$ and that $\Phi(G)$ is nilpotent.

To prove (iii), suppose that p does not divide $|G/\Phi(G)|$. Then P, a Sylow p-subgroup of G, is contained in $\Phi(G)$. By (ii), $\Phi(G)$ is nilpotent. Thus, by Lemma 1.46, P is normal in G.

Given that $P \lhd G$ with $(|P|, [G : P]) = 1$, Theorem 2.12 implies $G = [P]H$ for some subgroup H of G. This means

$$G = \langle P, H\rangle = \langle\Phi(G), H\rangle = \langle H\rangle = H,$$

which is a contradiction. Thus, p divides $|G/\Phi(G)|$. □

Lemma 1.52. *Let N be a normal subgroup of a group G. Then $\Phi(G)N/N \leq \Phi(G/N)$.*

Beweis. Let M/N be a maximal subgroup of G/N. This implies that M is a maximal subgroup of G. By Lemma 1.51, $\Phi(G) \leq M$ and $\Phi(G)N/N \leq M/N$. Applying Lemma 1.51 to G/N, it follows that $\Phi(G)N/N \leq \Phi(G/N)$. □

Theorem 1.53. *If N is a normal subgroup of a group G such that $N \leq \Phi(H)$ for some subgroup H of G, then $N \leq \Phi(G)$.*

Beweis. Suppose that $N \nleq \Phi(G)$. Then there is a maximal subgroup M of G such that $N \nleq M$ and $G = NM$. By the modular identity (Lemma 1.6),

$$H = G \cap H = NM \cap H = N(M \cap H).$$

Given that $N \leq \Phi(H)$, this implies $H = M \cap H$ or that $H \leq M$. Since $N \leq \Phi(H) \leq H$, it follows that $N \leq M$, a contradiction. Thus, N is contain in every maximal subgroup of G, and by Lemma 1.51, $N \leq \Phi(G)$. □

The following corollary follows immediately from this result.

Corollary 1.54. *If N is a normal subgroup of a group G, then $\Phi(N) \leq \Phi(G)$.*

Lemma 1.55. *If G_1 and G_2 are groups, then $\Phi(G_1 \times G_2) = \Phi(G_1) \times \Phi(G_2)$.*

Beweis. By Corollary 1.54, it follows that $\Phi(G_1)\times\Phi(G_2) \leq \Phi(G_1\times G_2)$. Now let M_i and M_j be a maximal subgroups of G_1 and G_2 respectively. Then $M_i\times G_2$ and $G_1\times M_j$ are maximal subgroups of $G_1 \times G_2$. Given that $\Phi(G_1\times G_2) \leq \bigcap_i M_i\times G_2$ and $\Phi(G_1\times G_2) \leq \bigcap_j G_1\times M_j$, it follows that $\Phi(G_1 \times G_2) = \Phi(G_1) \times \Phi(G_2)$. □

Lemma 1.56. *Let N and M be normal subgroups of a group G such that $M \leq N$ and $M \leq \Phi(G)$. If N/M is nilpotent, then N is nilpotent.*

Beweis. Let P be a Sylow p-subgroup of N. Then PM/M is a Sylow p-subgroup of N/M. Since N/M is nilpotent, Lemma 1.46 implies PM/M is characteristic in N/M. Given that $N \trianglelefteq G$, it follows that PM is normal in G.

Since P is a Sylow p-subgroup of PM, it follows by the Frattini argument (Lemma 1.7) that $G = PM\cdot\mathcal{N}_G(P) = M\mathcal{N}_G(P)$. However, $M \leq \Phi(G)$, which implies $G = \mathcal{N}_G(P)$ or that $P \trianglelefteq G$. Thus, $P \trianglelefteq N$ and N is nilpotent by Lemma 1.46. □

Corollary 1.57. *For a group G, if $G/\Phi(G)$ is nilpotent, then G is nilpotent.*

Theorem 1.58. *A group G is nilpotent if and only if $G' \leq \Phi(G)$.*

Beweis. ($\Rightarrow$) Assume G is nilpotent, and suppose $G' \nleq \Phi(G)$. Then there exists a maximal subgroup M of G such that $G' \nleq M$ and $G = G'M$.

Let

$$\{1\} = Z_0(G) \leq Z_1(G) \leq \cdots \leq Z_r(G) = G$$

be the upper central series for G. For each i, where $0 \leq i \leq r$, let $M_i = Z_i(G)M$. Let j be maximal such that $G = M_j$ and $M_{j-1} \lhd G$. Then G/M_{j-1} is abelian and $G' \leq M_{j-1} = Z_{j-1}(G)M$. This implies that $G = G'M \leq Z_{j-1}(G)M$, a contradiction. It now follows that $G' \leq \Phi(G)$.

($\Leftarrow$) Assume that $G' \leq \Phi(G)$. Then $G/\Phi(G)$ is abelian, and thus nilpotent by Theorem 1.46. By Corollary 1.57, G is nilpotent. □

There is a natural link between the Frattini subgroup and normal subgroup supplementation and complementation. While this relation will be explored in more detail in Chapter 3, a few simple observations are made here.

Given an arbitrary normal subgroup N of a group G, there always exists a subgroup H of G (take $H = G$), such that $G = NH$. If N is not contained in the Frattini subgroup of G, then there exists a proper subgroup K of G such that $G = NK$ (let K

be a maximal subgroup of G that does not contain N). This motivates the following definition and lemma.

Definition 1.59. For a normal subgroup N and a subgroup H of a group G, G is a **reduced product** of N by H if $G = NH$ and H does not contain a proper subgroup K such that $G = NK$.

Lemma 1.60. *If N is a normal subgroup of a group G, a reduced product $G = NH$ always exists with $N \cap H \leq \Phi(H)$.*

Beweis. If N is contained in $\Phi(G)$, the result is trivial. Assume that N is not contained in $\Phi(G)$. Then there exists a maximal subgroup M of G that does not contain N and $G = NM$. Thus, there exists a subgroup H contained in M such that $G = NH$ is a reduced product. Suppose that $N \cap H$ is not contained in $\Phi(H)$. Then there exists a proper subgroup K of H such that $H = \langle N \cap H, K\rangle$. Consequently, $G = \langle N, N \cap H, K\rangle = \langle N, K\rangle$. Since N is normal in G, we get $G = NK$, a contradiction. Thus, $N \cap H \leq \Phi(H)$. □

Now if N is an abelian minimal normal subgroup of a group G, then either $N \leq \Phi(G)$ or $N \cap \Phi(G) = \{1\}$. This motivates the following lemma.

Lemma 1.61. *Let N be an abelian minimal normal subgroup of a group G. If $N \nleq \Phi(G)$, then $G = [N]M$ for some maximal subgroup M in G.*

Beweis. Since $N \nleq \Phi(G)$, there must be a maximal subgroup M of G such that $N \nleq M$. As a result, $G = NM$. Given that N is abelian, $N \cap M$ is normal in G. This implies $N \cap M = \{1\}$ and that $G = [N]M$. □

Lemma 1.61, along with the fact that any chief factor H/K of a solvable group G is an abelian minimal normal subgroup of G/K, motivates the following definition.

Definition 1.62. Let H/K be a chief factor of a group G. H/K is a **Frattini chief factor** if $H/K \leq \Phi(G/K)$ and a **complemented chief factor** if H/K is complemented in G/K.

Lemma 1.63. *Let H/K be an abelian chief factor of a group G.*
(i) *H/K is either a Frattini or a complemented chief factor.*
(ii) *If H/K is a complemented chief factor, then for each complement M/K of H/K in G/K, M is a maximal subgroup of G.*

Beweis. For (i), suppose that $H/K \nleq \Phi(G/K)$, and assume without loss of generality that $K = \{1\}$. Then H is a minimal normal subgroup of G with $H \nleq \Phi(G)$. Thus, by Lemma 1.61, there is a maximal subgroup M of G such that $G = [H]M$.

To prove (ii), let L/K be a complement to H/K in G/K. Then $L \leq M$, where M is a maximal subgroup of G. Given that H is normal if G and not contained in M, $G = HM$ with $H/K \cap M/K \lhd G/K$. This implies $H/K \cap M/K = \{1_{G/K}\}$, and that M/K is a complement to H/K in G/K. Since $[G : M] = |H/K| = [G : L]$, L is maximal in G. □

Note that Lemma 1.63 implies that every chief factor in a solvable group is either a Frattini or complemented chief factor.

This treatise of the Frattini subgroup concludes with a brief discussion of the prefrattini subgroup of a solvable group, a concept introduced by Gaschütz [43]. Let G be a solvable group, and let H/K be a complemented chief factor of G with complement M/K. The subgroup M is maximal in G (Lemma 1.63). Given that $H/K \cap M/K = \{1_{G/K}\}$, it follows that $H/K \cap \mathrm{Core}_G(M)/K = \{1_{G/K}\}$ or that $\mathrm{Core}_G(M) \leq C_G(H/K)$. This leads to the following definition.

Definition 1.64. Let H/K be a complemented chief factor in a solvable group G. The **crown of G corresponding to H/K**, denoted by $Wr_G(H/K)$, is

$$Wr_G(H/K) = C_G(H/K)/C,$$

where $C = \bigcap_i \mathrm{Core}_G(M_i)$, where each M_i is a representative of a conjugacy class of maximal subgroups of G that complement H/K.

Each crown of a solvable group G corresponds to a complemented chief factor in G and each crown has a complement in G. Only the crowns of complemented chief factors that are non G-isomorphic need to be considered.

Definition 1.65. Let $Wr(G) = \{Wr_G(H_1/K_1), \ldots, Wr_G(H_n/K_n)\}$ be the crowns of non G-isomorphic complemented chief factors in a solvable group G with L_i the complement of $Wr_G(H_i/K_i)$. The subgroup $W = \bigcap_i L_i$ is a **prefrattini** subgroup of G.

Gaschütz developed prefrattini subgroups partial due to their following two properties which are presented without proof.

Theorem 1.66. [Satz 6.1 in Gaschütz [43]] *Let W be a prefrattini subgroup of a solvable group G. Then W avoids all complemented chief factors and covers all Frattini chief factors.*

Theorem 1.67. [Satz 6.5 in Gaschütz [43]] *The intersection of all of the prefrattini subgroups of a solvable group G is the Frattini subgroup of G.*

This section ends with a few comments concerning the Fitting subgroup. If N_1 and N_2 are normal nilpotent subgroups of a group G, then N_1N_2 is also a normal nilpotent subgroup of a group G. This motivates the following definition.

Definition 1.68. The maximal normal nilpotent subgroup of a group G is called the **Fitting subgroup** of G and is denoted by $\mathcal{F}(G)$.

Theorem 1.69. *Given a group G,*
(i) $\Phi(G) \leq \mathcal{F}(G)$,
(ii) *if G is solvable, then $C_G(\mathcal{F}(G)) \leq \mathcal{F}(G)$,*
(iii) *$\mathcal{F}(G)/\Phi(G) = \mathcal{F}(G/\Phi(G))$ and is the direct product of abelian minimal normal subgroups of $G/\Phi(G)$, and*
(iv) *if N is a minimal normal subgroup of G, then $\mathcal{F}(G) \leq C_G(N)$.*

Beweis. Item (i) follows directly from Lemma 1.51. To prove (ii), suppose that $C_G(\mathcal{F}(G)) \nleq \mathcal{F}(G)$. Then $N = C_G(\mathcal{F}(G)) \cap \mathcal{F}(G) < C_G(\mathcal{F}(G))$. Since $C_G(\mathcal{F}(G))$ is normal in G, let L be a normal subgroup of G such that $L \leq C_G(\mathcal{F}(G))$ and L is minimal such that $N < L$. Thus, L/N is a minimal normal subgroup of G/N and is elementary abelian. This means $[L, L] \leq N$. But given that $L \leq C_G(\mathcal{F}(G))$ and $N \leq \mathcal{F}(G)$, it follows that $[[L, L], L] = \{1\}$. Thus, L is nilpotent and $L \leq C_G(\mathcal{F}(G)) \cap \mathcal{F}(G)$, a contradiction. This implies $C_G(\mathcal{F}(G)) \leq \mathcal{F}(G)$.

To prove the first part of (iii), first note that $\mathcal{F}(G)/\Phi(G) \leq \mathcal{F}(G/\Phi(G))$. Now let $\mathcal{F}(G/\Phi(G)) = N/\Phi(G)$. Given that $N/\Phi(G)$ is nilpotent, Lemma 1.56 implies that $N \leq \mathcal{F}(G)$ and that $\mathcal{F}(G/\Phi(G)) \leq \mathcal{F}(G)/\Phi(G)$.

For the second part of (iii), proceed by assuming $\Phi(G) = \{1\}$. By Corollary 1.54 and Lemma 1.58,

$$\mathcal{F}(G)' \leq \Phi(\mathcal{F}(G)) \leq \Phi(G) = \{1\}.$$

This means $\mathcal{F}(G)$ is abelian. Now let T be the product of the abelian minimal normal subgroups of G. If follows that $T \leq \mathcal{F}(G)$. Assume that $T < \mathcal{F}(G)$. Using Theorem 3.9, $G = [T]H$ for some subgroup H of G. By Lemma 1.11, it follows that

$$\mathcal{F}(G) = T(H \cap \mathcal{F}(G)) = T \times (H \cap \mathcal{F}(G)),$$

where $H \cap \mathcal{F}(G) \trianglelefteq G$. If $H \cap \mathcal{F}(G) \neq \{1\}$, then there exists a abelian minimal normal subgroup of G not in T. This is a contradiction. Thus, $T = \mathcal{F}(G)$ and $\mathcal{F}(G)$ and is the direct product of abelian minimal normal subgroups of G.

For (iv), first consider the case that $N \cap \mathcal{F}(G) = \{1\}$. Since both are normal in G, $[N, \mathcal{F}(G)] = \{1\}$ and $\mathcal{F}(G) \leq C_G(N)$. If this is not the case, then $N \leq \mathcal{F}(G)$. Given that $\mathcal{F}(G)$ is nilpotent, $N \cap Z(\mathcal{F}(G)) \neq \{1\}$. However, $N \cap Z(\mathcal{F}(G)) \lhd G$, which means $N \leq Z(\mathcal{F}(G))$ and that $\mathcal{F}(G) \leq C_G(N)$. □

Lemma 1.70. *Let G be a solvable group. If N is the unique minimal normal subgroup of G, such that $N \nleq \Phi(G)$, then $N = C_G(N) = \mathcal{F}(G)$.*

Beweis. If G is nilpotent, then $N \cap Z(G) \neq \{1\}$ by Lemma 1.45. This implies that $N \leq Z(G)$ and by Lemma 1.61 that $G = N \times H$, where H is a maximal subgroup of G. This contradictions the fact the N is unique. Thus, G is not nilpotent.

By Theorem 1.69, $N \leq \mathcal{F}(G) \leq C_G(N)$. Given that $N \nleq \Phi(G)$, by Lemma 1.61 there is a maximal subgroup M of G such that $G = NM$ with $N \cap M = \{1\}$. Since $C_G(N) \lhd G$, it follows that $C_G(N) \cap M \lhd M$. This implies that $C_G(N) \cap M \lhd G$, which forces $C_G(N) \cap M = \{1\}$. By the modular identity (Lemma 1.6)

$$C_G(N) = G \cap C_G(N) = NM \cap C_G(N) = N(M \cap C_G(N)) = N.$$

Since $\mathcal{F}(G) \leq C_G(N)$, it follows that $N = C_G(N) = \mathcal{F}(G)$. □

1.7 Permutable and modular subgroups

Definition 1.71. Let $\mathcal{X}$ be a family of subgroups of a group G. A subgroup A of G is **$\mathcal{X}$-permutable** if A permutes with all subgroups B in $\mathcal{X}$.

Let A be subgroup of a group G that is $\mathcal{X}$-permutable in G. If $\mathcal{X}$ is the collection of all subgroups of G, then A is simply said to be **permutable** in G. Permutable subgroups, originally called quasinormal subgroups, were introduced by Ore in [86]. The subgroup A is called **S-permutable** in G when $\mathcal{X}$ is the family of Sylow subgroups of G and is called **M-permutable** when $\mathcal{X}$ is the family of maximal subgroups of G. For more information of $\mathcal{X}$-permutability see [3].

The concept of a modular element, originally called a Dedekind element, was introduced by Kurosh [75].

Definition 1.72. A subgroup M of a group G is a **modular subgroup** if the following two conditions are satisfied.
(i) $\langle A, M\rangle \cap B = \langle A, M \cap B\rangle$ for all $A, B \leq G$ such that $A \leq B$.
(ii) $\langle A, M\rangle \cap B = \langle A \cap B, M\rangle$ for all $A, B \leq G$ such that $M \leq B$.

The group G is **modular** or has a **modular subgroup lattice** if every subgroup of the group is a modular subgroup.

Clearly, all normal subgroups of a group are modular subgroups and permutable in the group.

Theorem 1.73. *A subgroup M of a finite group is permutable in G if and only if M is modular and subnormal in G.*

The proof given here is due to Schmidt [92].

Beweis. First suppose that M is permutable in G, and let $B, C \leq G$ such that $B \leq C$. Clearly, $\langle B, M \cap C\rangle \leq \langle B, M\rangle \cap C$. Let $x \in \langle B, M\rangle \cap C$. Since $\langle B, M\rangle = BM$, the element $x = bm$ for $b \in B$ and $m \in M$. Then $m = b^{-1}x$ and $m \in C$. Thus, $x = bm$ for $b \in B$ and $m \in M \cap C$. It follows that $x \in \langle B, M \cap C\rangle$ and that $\langle B, M\rangle \cap C \leq \langle B, M \cap C\rangle$.

Now let $X, Y \leq G$ such that $M \leq Y$. The fact that $\langle X, M\rangle \cap Y = \langle X \cap Y, M\rangle$ follows directly from the modular identity (Lemma 1.6). Thus, M is modular in G.

To show that M is subnormal in G, proceed by induction on the order of G. Let N be a maximal permutable subgroup of G that contains M. Suppose that N is not normal in G. Then there is an element $g \in G$ such that $N^g \neq N$. Let H be an arbitrary subgroup of G. Then

$$NN^gH = Ng^{-1}NH^{g^{-1}}g = Ng^{-1}H^{g^{-1}}Ng = NHN^g = HNN^g,$$

which implies that NN^g is permutable in G. Thus, $G = NN^g$ and $g = xy$ for $x \in N$ and $y \in N^g$. As a result $N = N^x = N^{gy^{-1}} = N^g$, which is a contradiction. Consequently N is normal in G and by induction M is subnormal in N. This implies that M is subnormal in G.

Now assume that M is a modular and subnormal subgroup of G. Proceed by induction on the order of G. Let H be an arbitrary subgroup of G. If $H \leq M$, then trivially $HM = MH$. If $H \nleq M$, then $M < \langle H, M\rangle$. Given that M is also modular and subnormal in $\langle H, M\rangle$, there exists a proper normal subgroup N of $\langle H, M\rangle$ such that $M \leq N < G$. By induction, $M(H\cap N) = (H\cap N)M$. Since M is modular in G, the subgroup $N = \langle H, M\rangle \cap N = \langle H \cap N, M\rangle$ and $N = (H \cap N)M$. Thus,

$$HM = H(H \cap N)M = HN = NH = M(H \cap N)H = MH.$$

Consequently M is permutable in G. □

Given that in a finite p-group every subgroup is subnormal, the follow corollary follows directly from Theorem 1.73

Corollary 1.74. *A finite p-group has a modular subgroup lattice if and only if any two of its subgroups permute.*

For more information of modular and permutable subgroups, see [92].

1.8 Wreath products

This section presents concepts and results central to the early development of normal subgroup complementation. The treatise given here very closely follows the exposition by Bechtell in his book *Theory of Groups* [7].

Let G be a nontrivial group, let I be a finite index set (for example $I = \{1, 2, \dots, n\}$), and let A be a subgroup of $\mathcal{S}_I$, the collection of permutations of I (if $I = \{1, 2, \dots, n\}$, then $\mathcal{S}_I = \mathcal{S}_n$). Define $G_I = \prod_{i \in I} G_i$, where $G_i \cong G$ for each $i \in I$ ($G_I = G_1 \times \cdots \times G_n$ if $I = \{1, 2, \dots, n\}$).

Given that I is any index set, it is more convenient to identify each element of G_I as a mapping $f : I \to G$. For $f, g \in G_I$, define $fg(i) = f(i)g(i)$ for $i \in I$. With this operation on G_I, the identity element id in G_I is the mapping from I to G such that $id(i) = e$ for all $i \in I$. Given $f \in G_I$, the inverse of f or f^{-1} is defined as $f^{-1}(i) = (f(i))^{-1}$ for each $i \in I$. The collection of elements G_i with this operation is a group.

Now consider the subgroup A of the permutation group S_I. The goal is to create a monomorphism θ from A into $Aut(G_I)$. Let $\alpha \in A$, and let $f \in G_I$. Define the element $f^{\overline{\alpha}}$ in $Aut(G_I)$ by $f^{\overline{\alpha}}(i) = f(\alpha(i))$ for all $i \in I$. Thus, for each $\alpha \in A$, $\overline{\alpha}$ defines the mapping $\overline{\alpha}$ from G_I to G_I by $\overline{\alpha}(f) = f^{\overline{\alpha}}$ for all $f \in G_I$. Let $f, g \in G_I$ and suppose $\overline{\alpha}(f) = \overline{\alpha}(g)$. Then $f^{\overline{\alpha}}(i) = g^{\overline{\alpha}}(i)$ or $f(\alpha(i)) = g(\alpha(i))$ for all $i \in I$. Since α is a permutation of I, this implies $f(j) = g(j)$ for all $j \in I$ or that $f = g$. Thus, $\overline{\alpha}$ is injective.

Now let $f \in G_I$ and consider the element $f^{\overline{\alpha^{-1}}}$. Then $\overline{\alpha}(f^{\overline{\alpha^{-1}}}) = (f^{\overline{\alpha^{-1}}})^{\overline{\alpha}}$, where $(f^{\overline{\alpha^{-1}}})^{\overline{\alpha}}(i) = (f^{\overline{\alpha^{-1}}})(\alpha(i)) = f(\alpha^{-1}(\alpha(i))) = f(i)$ for all $i \in I$. This implies $\overline{\alpha}(f^{\overline{\alpha^{-1}}}) = f$ and that $\overline{\alpha}$ is surjective. Finally, the mapping $\overline{\alpha}$ is a homomorphism. Let $f, g \in G_I$ and consider $\overline{\alpha}(fg)$. Since

$$(fg)^{\overline{\alpha}}(i) = fg(\alpha(i)) = f(\alpha(i))g(\alpha(i)) = f^{\overline{\alpha}}(i)g^{\overline{\alpha}}(i)$$

for all $i \in I$, it follows that $\overline{\alpha}(fg) = \overline{\alpha}(f)\overline{\alpha}(g)$ and α is a homomorphism. As a result, $\overline{\alpha} \in Aut(G_I)$.

Now consider $\alpha, \beta \in A$. Given that

$$f^{\overline{\alpha\beta}}(i) = f(\alpha\beta(i)) = f(\alpha(\beta(i))) = f^{\overline{\alpha}}(\beta(i)) = f^{\overline{\alpha}\overline{\beta}}(i)$$

for all $i \in I$, it follows that $\overline{\alpha\beta} = \overline{\alpha}\overline{\beta}$ and that the mapping $\theta : A \to Aut(G_I)$ defined by $\theta(\alpha) = \overline{\alpha}$ is a homomorphism. Now let $\gamma \in ker(\theta)$. Then $f^{\overline{\gamma}} = f$ for all $f \in G_I$. This implies that $f(i) = f^{\overline{\gamma}}(i) = f(\gamma(i))$ for all $i \in I$. If γ is nontrivial, this can only occur if G is the trivial group. This contradiction implies that $\gamma = id$ and that the mapping θ is injective. Thus, the mapping $\theta : A \to Aut(G_I)$ is a monomorphism.

Definition 1.75. The **wreath product** of a nontrivial group G with respect to a finite set I and a subgroup A of S_I is $[G_I]_\theta A$, where θ is the monomorphism from A into $Aut(G_I)$ defined above. This semidirect product is denoted by $G\,wr\,A$. Given a group H, let L_H denote the left Cayley group on H. The wreath product created by taking $I = H$ and $A = L_H$ is the **standard wreath product** and is denoted by $G \wr H$.

For groups G and H, $G \wr H = [G_H]_\theta L_H$. Each permutation $l \in L_H$ is induced by left multiplication by an element $k^{-1} \in H$. In this way, l is associated with k and $f^k(h) =$

$f(k^{-1}h)$ for all $h \in H$ and $f \in G_H$. Thus, $G \wr H$ is the collection of elements $[G_H]_\theta H$, where for $f, g \in G_H$ and $r, s \in H$, $(fr)(gs) = fg^r rs$. The element fg^r satisfies $fg^r(h) = f(h)g^r(h) = f(h)g(r^{-1}h)$ for all $h \in H$.

Theorem 1.76 (Theorem 3.4.2 in Bechtell [7]). *For each normal subgroup N of a group G, there is a monomorphism from G into N wr G/N.*

Bechtell's proof. Let $H = G/N$. Then for each $h \in H$, let $h^\tau N$ denote the left coset of N in G corresponding to h. Furthermore, for each $g \in G$ let h_g denote the left coset in H that contains g. This implies that

$$h^\tau N = h = (h_g h_g^{-1})h = h_g(h_g^{-1}h)^\tau N = gN(h_g^{-1}h)^\tau N = g(h_g^{-1}h)^\tau N.$$

Thus, for each $g \in G$, there exists an element $n_{h,g} \in N$ such that $n_{h,g} = (h^\tau)^{-1}g(h_g^{-1}h)^\tau$. Define the mapping $\alpha : G \to N\,wr\,H$ by $\alpha(g) = \omega h_g$, where $\omega(h) = n_{h,g}$ for each $h \in H$.

Consider $f \in G$ and let $\alpha(f) = \mu h_f$, where $\mu(h) = n_{h,f}$ for each $h \in H$. Now

$$\alpha(f)\alpha(g) = \mu h_f \omega h_g = \mu\omega^{h_f} h_f h_g,$$

where

$$\mu\omega^{h_f}(h) = \mu(h)\omega^{h_f}(h) = \mu(h)\omega(h_f^{-1}h) = n_{h,f}n_{h_f^{-1}h,g}.$$

Considering $\alpha(fg) = \lambda h_{fg}$, where $\lambda(h) = n_{h,fg} = (h^\tau)^{-1}fg(h_{fg}^{-1}h)^\tau$. This implies

$$(h^\tau)^{-1}fg = n_{h,fg}((h_{fg}^{-1}h)^\tau)^{-1} = n_{h,fg}((h_g^{-1}h_f^{-1}h)^\tau)^{-1}.$$

At the same time,

$$\begin{aligned}(h^\tau)^{-1}fg &= ((h^\tau)^{-1}f)g \\ &= ((h^\tau)^{-1}f(h_f^{-1}h)^\tau((h_f^{-1}h)^\tau)^{-1})g \\ &= n_{h,f}(((h_f^{-1}h)^\tau)^{-1})g \\ &= n_{h,f}((h_f^{-1}h)^\tau)^{-1}g((h_g^{-1}h_f^{-1}h)^\tau)((h_g^{-1}h_f^{-1}h)^\tau)^{-1} \\ &= n_{h,f}((h_f^{-1}h)^\tau)^{-1}g((h_f^{-1}h)_g h_f^{-1}h)^\tau((h_g^{-1}h_f^{-1}h)^\tau)^{-1} \\ &= n_{h,f}n_{h_f^{-1}h,g}((h_g^{-1}h_f^{-1}h)^\tau)^{-1}.\end{aligned}$$

Thus, $n_{h,fg} = n_{h,f}n_{h_f^{-1}h,g}$ and $\alpha(fg) = \alpha(f)\alpha(g)$. This means that α is a homomorphism.

Now suppose that $\alpha(g) = \omega h_g$ is the identity element of $N\,wr\,H$. Then for all $h \in H$

$$\omega(h) = (h^\tau)^{-1}g(h_g^{-1}h)^\tau = (h^\tau)^{-1}g(1_H h)^\tau = (h^\tau)^{-1}gh^\tau = id.$$

This means $g = 1$ and that α is a monomorphism. □

Corollary 1.77 (Corollary 3.4.2 in Bechtell [7]). *For each normal subgroup N of a group G, there exists a group $\widetilde{G} = [\widetilde{N}]\,\widetilde{H}$ containing G such that $\widetilde{N} \cap G = N$ and $\widetilde{H} \cong G/N$.*

Bechtell's proof. By Theorem 1.76, there is a monomorphism $\alpha : G \to N\,wr\,G/N$ such that $\widehat{G} = N\,wr\,G/N = [N_H]_\theta G/N$, where θ is the monomorphism from G/N into $Aut(N_H)$ defined in Definition 1.75. The subgroup $\widehat{N} \cong N_H$ of $\widehat{G}$, defined by $\widehat{N} = \{\mu \cdot 1_{G/N} \mid \mu \in N_H\}$ is a normal subgroup of $\widehat{G}$. The subgroup $\widehat{H}$ of $\widehat{G}$ defined by $\widehat{H} = \{1 \cdot gN \mid gN \in G/N\}$ is isomorphic to G/N. By the definition of $\widehat{G}$, we have $\widehat{G} = \widehat{N}\widehat{H}$ with $\widehat{N} \cap \widehat{H} = 1 \cdot 1_{G/N} = 1$. Defining the mapping $\alpha : G \to \widehat{G}$ as presented in Theorem 1.76, $\alpha(N) = \alpha(G) \cap \widehat{N}$. By Theorem 2.2.5 in [7], there is a group $\widetilde{G}$ such that $G \subseteq \widetilde{G}$ and an isomorphism $\beta : \widetilde{G} \to \widehat{G}$, where $\beta|_G = \alpha$. Set $\widetilde{N} = \beta^{-1}(\widehat{N})$ and $\widetilde{H} = \beta^{-1}(\widehat{H})$. Then $\widetilde{G} = \widetilde{N}\widetilde{H}$ with $\widetilde{N} \cap \widetilde{H} = \{1\}$ and $N = \widetilde{N} \cap G$. □

Lemma 1.78 (Lemma 4.1.2 in Bechtell [7]). *Consider a group G having an abelian normal π-subgroup N such that for subgroups A and B of G, $G = [N]B = NA$. If $k = [B : A \cap B]$ is a π'-number, then for some $y \in N$, $B^y \subseteq A$.*

Bechtell's proof. First consider the case that $N \cap A = \{1\}$. Then $A \cong G/N \cong B$ and A and B are isomorphic. Define the isomorphic mapping $\alpha : B \to A$, where $\alpha(b)$ satisfies $bN = \alpha(b)N$ for each $b \in B$. Let L be the complete set of left coset representatives of $A \cap B$ in B. The order of L is k. Now for all $b \in B$, $\alpha(b)^{-1}b \in N$, and given that N is abelian, the element

$$z = \prod_L \alpha(b)^{-1}b$$

is well-defined. Given an element $a \in A \cap B$, then $\alpha(a) = a$ and

$$\alpha(ab)^{-1}(ab) = (\alpha(a)\alpha(b))^{-1}ab = \alpha(b)^{-1}\alpha(a)^{-1}ab = (\alpha(b))^{-1}a^{-1}ab = \alpha(b)^{-1}b$$

and z is independent of the coset representatives chosen.

Furthermore, for each element $d \in B$, the set $Ld = \{ld \mid l \in L\}$ is also a complete set of coset representatives for $A \cap B$ in B. It then follows that

$$\begin{aligned}\alpha(d)^{-1}z\alpha(d) &= \alpha(d)^{-1}\left[\prod_L \alpha(b)^{-1}b\right]\alpha(d)\\ &= \prod_L \alpha(d)^{-1}\alpha(b)^{-1}b\alpha(d)\\ &= \prod_L (\alpha(b)\alpha(d))^{-1}b\alpha(d)\end{aligned}$$

$$= \prod_{L} \alpha(bd)^{-1} b\alpha(d)$$
$$= \prod_{L} \alpha(bd)^{-1} bdd^{-1}\alpha(d)$$
$$= \left[\prod_{Ld} \alpha(bd)^{-1} bd\right] [d^{-1}\alpha(d)]^k$$
$$= z[d^{-1}\alpha(d)]^k.$$

Given that N is an abelian π-group with $k \in \pi'$, $N^k = \{n^k \mid n \in N\} = N$ and the mapping $\beta : N \to N$ defined by $\beta(n) = n^k$ is an automorphism of N. Thus, for some $x \in N$, $x^k = z$ and $\alpha(d)^{-1}x^k\alpha(d) = [\alpha(d)^{-1}x\alpha(d)]^k$. The previous calculation showed that $\alpha(d)^{-1}x^k\alpha(d) = x^k[d^{-1}\alpha(d)]^k$. This implies that

$$[\alpha(d)^{-1}x\alpha(d)]^k = x^k[d^{-1}\alpha(d)]^k = [xd^{-1}\alpha(d)]^k.$$

or that $\alpha(d)^{-1}x\alpha(d) = xd^{-1}\alpha(d)$. This means $\alpha(d)^{-1}x = xd^{-1}$ or that $\alpha(d) = xdx^{-1}$ for all $d \in B$. This means $B^y \subseteq A$ for $y = x^{-1}$.

Now suppose $N \cap A \neq \{1\}$. Since N is abelian, $N \cap A = M \lhd G$. Thus, N/M is an abelian normal π-subgroup of G/M. Furthermore, by Lemma 1.11, $G/M = [N/M]BM/M = N/M \cdot A/M$. Since $BM/M \cong B/(B \cap M) \cong B$ and

$$BM/M \cap A/M = (BM \cap A)/M = M(B \cap A)/M \cong (B \cap A)/(A \cap B \cap M) \cong B \cap A$$

it follows that $[BM/M : BM/M \cap A/M] = [B : A \cap B] = k$, which is a π'-number. Given that $BM/M \cap AM/M = \{1_{G/M}\}$, we know from the first part of the proof that there is an element $yM \in N/M$ such that $(yM)^{-1}BM/M(yM) \subseteq AM/M$ or that $y^{-1}ByM/M \subseteq A/M$. As a result, $B^y \subseteq (BM)^y \subseteq A$. ☐

Lemma 1.79 (Lemma 4.1.3 in Bechtell [7]). *Let $N \operatorname{wr} H = [\widetilde{N}]\widetilde{H}$, where $\widetilde{N} = \{f \cdot 1_H \mid f \in N_H\} \cong N_H$ and $\widetilde{H} = \{1_{N_H} \cdot h \mid h \in H\} \cong H$. If a subgroup C of $N \operatorname{wr} H$ contains $\widetilde{N}$ and $C = [\widetilde{N}]A = [\widetilde{N}]B$, then there exists an element $x \in \widetilde{N}$ such that $A^x = B$.*

Beweis. By the modular identity (Lemma 1.6) and Lemma 1.11,

$$C = \widetilde{N}\widetilde{H} \cap C = \widetilde{N}(\widetilde{H} \cap C) = \widetilde{N}(C \cap \widetilde{H}) = [\widetilde{N}]\widetilde{H}_1,$$

where $\widetilde{H}_1 = C \cap \widetilde{H}$. Let $H_1 = \{h \in H \mid 1 \cdot h \in \widetilde{H}_1\}$. Each of the subgroups A, B, and $\widetilde{H}_1$ gives a complete set of coset representatives for $\widetilde{N}$ in C. Thus, for each $a \in A$, there is an element $1_{N_H}h \in \widetilde{H}_1$ such that $a\widetilde{N} = 1_{N_H}h\widetilde{N}$. Consequently $a(1_{N_H}h)^{-1} \in \widetilde{N}$ and $a(1_{N_H}h)^{-1} = f_h \cdot 1_H$, where $f_h \in N_H$. This implies $a = (f_h 1_H)(1_{N_H}h) = f_h h$. In the same manner, each element $b \in B$ has a unique expression of the form $g_h h$ for $g_h \in N_H$ and $h \in \overline{H}_1$.

For two elements $a = f_h h$ and $a' = f_k k$ in A,

$$aa' = f_h h f_k k = f_h f_k^h hk \in A,$$

where $aa' = f_{hk} hk$. This implies that $f_h f_k^h = f_{hk}$ or that $f_k^h = f_k^{-1} f_{hk}$. Again, in a similar manner, $g_{hk} = g_h g_k^h$. Let $\mathcal{H}$ be the set of right cosets representatives of H_1 in H. Define the mapping $\alpha : H \to N$ by $\alpha(hx) = f_h(hx)g^{-1}(hx)$ for $h \in H_1$ and $x \in \mathcal{H}$. It then follows that

$$\begin{aligned}
f_k(hx)\alpha^k(hx) &= f_k(hx)\alpha(k^{-1}hx) \\
&= f_k(hx)f_{k^{-1}h}(k^{-1}hx)g_{k^{-1}h}^{-1}(k^{-1}hx) \\
&= f_k(hx)f_{k^{-1}h}(k^{-1}hx)(g_{k^{-1}h}^{k})^{-1}(hx) \\
&= f_k(hx)f_{k^{-1}h}^{k})(hx)(g_{k^{-1}h}^{k})^{-1}(hx) \\
&= f_k(hx)(f_k^{-1}f_{kk^{-1}h})(hx)(g_k^{-1}g_{kk^{-1}h})^{-1}(hx) \\
&= f_k(hx)(f_k^{-1}f_h)(hx)(g_h^{-1}g_k)(hx) \\
&= f_h(hx)g_h^{-1}(hx)g_k(hx) \\
&= \alpha(hx)g_k(hx).
\end{aligned}$$

This implies $f_k\alpha^k = \alpha g_k$. Thus, for all $k \in H_1$,

$$(f_k k)(\alpha 1) = f_k\alpha^k k \cdot 1 = f_k\alpha^k k = \alpha g_k k = \alpha g_k^1 k = (\alpha 1)(g_k k).$$

Therefore $(\alpha 1)^{-1}(f_k k)(\alpha 1) = g_k k$, which implies $A^x = B$ for $x = \alpha 1$ □

Theorem 1.80 (Theorem 4.1.4 in Bechtell [7]). *Consider a group G having a subgroup B such that B contains an abelian π-subgroup N, where N is normal in G and $[G : B] = k$ is a π'-number.*
(i) *If B splits over N, then G splits over N.*
(ii) *The number of distinct conjugacy classes of the complements of N in G does not exceed the number of distinct conjugacy classes of the complements of N in B.*

Bechtell's proof. For (i), let $H = G/N$ and consider the wreath product $K = N \,wr\, H$. By Corollary 1.77, G is contained in $K = [N_H]H$ such that $N_H \cap G = N$. Clearly, N_H is a π-group where $K = N_H G$ with $N \le B \cap N_H \le G \cap N_H = N$ or $B \cap N_H = N$. This implies $[K : N_H B] = k$.

Given that B splits over N, $B = [N]C$ for some subgroup $C \le B$. By Lemma 1.11, this implies $N_H B = [N_H]C$. Consequently, $N_H B = N_H H \cap N_H B = N_H(H \cap N_H B)$ and

$N_H \cap (H \cap N_H B) = (N_H \cap H) \cap N_H B = \{1\}$. This means $N_H B = [N_H](H \cap N_H B)$. By Lemma 1.79, there is an element $x \in N_H$ such that $C^x = H \cap N_H B \subseteq H$.

Note that

$$k = [K : N_H B] = [H : H \cap N_H B] = [H : C^x] = [H : G^x \cap H][G^x \cap H : C^x],$$

and let $l = [H : G^x \cap H]$. Then $l|k$ and l is a π'-number. Given that $K = N_H G^x$, Lemma 1.78 implies there is an element $y \in N_H$ such that $H^y \subseteq G^x$. As a result, $H^{yx^{-1}} \subseteq G$, where $yx^{-1} \in N_H$. Thus, $G = NH^{yx^{-1}}$ and $N \cap H^{yx^{-1}} \subseteq N_H \cap H^{yx^{-1}} = \{1\}$. Consequently $G = [N]H^{yx^{-1}}$ and G splits over N.

To prove (ii), suppose $G = [N]C = [N]D$ for subgroups C and D of G. Then by Lemma 1.11, $B = [N](C \cap B) = [N](D \cap B)$. Suppose there exists an element $b \in B$ such that $(C \cap B) = (D \cap B)^b$. Given that $(D \cap B)^b \subseteq D^b \cap B$ and that $B \cap C \subseteq C$, it follows that $C \cap B \subseteq C \cap D^b$. Now given that $G = NC$ with $N \leq B$, $G = BC$ and B and C permute with $[BC : B] = [C : C \cap B]$. This implies

$$k = [G : B] = [BC : B] = [C : C \cap B] = [C : C \cap D^b][C \cap D^b : C \cap B]$$

and that $m = [C : C \cap D^b]$ divides k. Thus, m is a π'-number and by Lemma 1.78 C is conjugate to D^b in G or that C and Dc are conjugate in G. Thus, when $B \cap C$ and $B \cap D$ are conjugate in B, C and D are conjugate in G □

1.9 Formations

This section provides a brief introduction to formations. The investigation of formations presented here is rooted in the chapter "Classes of Finite Solvable Groups", written by Gary Walls, from the book *Between Nilpotent and Solvable* [106], and many of the results and proof structures in this section stem from it.

Definition 1.81. A **formation** $\mathfrak{F}$ is a collection of solvable groups with the following properties:
(i) If $G \in \mathfrak{F}$ and $N \trianglelefteq G$, then $G/N \in \mathfrak{F}$.
(ii) If $N_1, N_2 \trianglelefteq G$ such that $G/N_1, G/N_2 \in \mathfrak{F}$, then $G/N_1 \cap N_2 \in \mathfrak{F}$.

The collection $\mathfrak{A}$ of abelian groups is a formation. This follows from the fact that for a group G and a normal subgroup N of G, G/N is abelian if and only if $G' \leq N$. The collection $\mathfrak{N}$ of nilpotent groups is a formation by Lemma 1.45 and the fact that $G/N_1 \cap N_2$ is isomorphic to a subgroup of $G/N_1 \times G/N_2$. The collection $\mathfrak{S}$ of solvable groups, the collection $\mathfrak{U}$ of supersolvable groups, and for π a set of primes, the collection $\mathfrak{S}_\pi$ of π-groups are formations.

Definition 1.82. A nonempty formation $\mathfrak{F}$ is **normal** if and only if group $G \in \mathfrak{F}$ implies that each normal subgroup of G is also in $\mathfrak{F}$.

Each of the formations listed above are normal formations. The following definition is motivated by the fact that if $G/\Phi(G)$ is nilpotent for some group G, then G is nilpotent (Corollary 1.57).

Definition 1.83. A nonempty formation $\mathfrak{F}$ is **saturated** if the following condition is satisfied: If G is a group with $G/\Phi(G) \in \mathfrak{F}$, then $G \in \mathfrak{F}$.

The formation $\mathfrak{A}$ of abelian groups is not saturated as demonstrated by any nonabelian p-group. Formations $\mathfrak{N}$, $\mathfrak{S}$, $\mathfrak{U}$, and $\mathfrak{G}_\pi$ are all saturated.

Theorem 1.84. [Gaschütz and Lubeseder [45, 78]] *A formation $\mathfrak{F}$ is saturated if and only if the following condition is satisfied: If a group $G \notin \mathfrak{F}$ and N is a minimal normal subgroup of G with $G/N \in \mathfrak{F}$, then N has a complement in G and all complements of N in G are conjugate in G.*

Beweis. ($\Rightarrow$) Let G be a group such that $G \notin \mathfrak{F}$, and let N be a minimal normal subgroup of G such that $G/N \in \mathfrak{F}$. There are two cases to examine.

Case 1: $\Phi(G) \neq \{1\}$. Proceed by induction of the order of G. If $N \leq \Phi(G)$, then $\Phi(G)/N \leq G/N$, which implies $(G/N)/(\Phi(G)/N) \cong G/\Phi(G)$ is in $\mathfrak{F}$. Since $\mathfrak{F}$ is saturated, it follows that $G \in \mathfrak{F}$, a contradiction. Thus, $N \nleq \Phi(G)$, which implies $N \cap \Phi(G) = \{1\}$.

Given that $N \cap \Phi(G) = \{1\}$, $\Phi(G)N/\Phi(G)$ is a minimal normal subgroup of $G/\Phi(G)$. Note that $G/\Phi(G) \notin \mathfrak{F}$ since $G \notin \mathfrak{F}$. It then follows by Lemma 1.61 that $\Phi(G)N/\Phi(G)$ has a complement $M/\Phi(G)$ in $G/\Phi(G)$, where $M/\Phi(G)$ is maximal in $G/\Phi(G)$. This means $G = \Phi(G)NM = NM$. Since N is minimal normal (and abelian) in G, it follows that $N \cap M = \{1\}$ and that $G = [N]M$.

Now let L be another complement to N in G. Then L is maximal in G, and by Lemma 1.11, $L/\Phi(G)$ and $M/\Phi(G)$ are complements to $\Phi(G)N/\Phi(G)$ in $G/\Phi(G)$. By induction $L/\Phi(G)$ and $M/\Phi(G)$ are conjugate in $G/\Phi(G)$. Thus, by Lemma 1.14, L and M are conjugate in G.

Case 2: $\Phi(G) = \{1\}$. Proceed by induction on the order of G. By Lemma 1.61, there exists a maximal subgroup M of G such that $G = [N]M$. Let L be another complement to N in G.

Suppose there exists another minimal normal subgroup K in G. Then $G/K \notin \mathfrak{F}$, since $G/K \in \mathfrak{F}$ would imply that $G/N \cap K = G \in \mathfrak{F}$, a contradiction. If $K \nleq M$, then $G = [K]M$ and $G/K \cong M \cong G/N \in \mathfrak{F}$, another contradiction. Thus, $K \leq M$, and by a similar argument, $K \leq L$. By Lemma 1.11,

$$G/K = [KN/K]M/K = [KN/K]L/K.$$

Thus, by induction, M/K and L/K are conjugate in G/K. Thus, by Lemma 1.14, M and L are conjugate in G.

Now assume that N is the unique minimal normal subgroup of G. By Lemma 1.70, $C_G(N) = N$, where N is an abelian p-group for some prime p. Let Q/N be minimal normal in G/N. Since $C_G(N) = N$, it follows that $|Q/N| = q^\alpha$ for some prime $q \neq p$.

Given that $G = [N]M = [N]L$, Lemma 1.11 implies that $Q = [N](M \cap Q) = [N](L \cap Q)$, where $M \cap Q$ and $L \cap Q$ are Sylow q-subgroups of Q. Thus, there exists an element $x \in Q$ such that $L \cap Q = (M \cap Q)^x$ or $L \cap Q = M^x \cap Q$. Since Q is normal in G, $L \cap Q \trianglelefteq L$ and $M \cap Q \trianglelefteq M$. This implies $L \cap Q = (M \cap Q)^x \trianglelefteq M^x$. Thus, $L \leq N_G(L \cap Q)$ and $M^x \leq N_G(L \cap Q)$. But given that $L \cap Q$ is not normal in G (recall that N is the unique minimal normal subgroup of G), $\langle L, M^x \rangle < G$. Since L and M^x are maximal in G, this implies $L = M^x$.

($\Leftarrow$) Let G be a group such that $G/\Phi(G) \in \mathfrak{F}$. If $\Phi(G) = \{1\}$, then trivially $G \in \mathfrak{F}$. Assume that $\Phi(G) \neq \{1\}$ and that $G \notin \mathfrak{F}$. Proceed by induction on the order of G. Let N be a minimal normal subgroup of G such that $N \leq \Phi(G)$. It then follows by Lemma 1.52 that $\Phi(G/N) = \Phi(G)/N$.

Consider the group G/N. Given that

$$(G/N)/\Phi(G/N) = (G/N)/(\Phi(G)/N) \cong G/\Phi(G) \in \mathfrak{F},$$

it follows by induction that $G/N \in \mathfrak{F}$. But since $G \notin \mathfrak{F}$, there is a maximal subgroup M of G such that $G = NM$. This is a contradiction as $N \leq \Phi(G) \leq M$. Thus, $G \in \mathfrak{F}$. □

Definition 1.85. Let $\mathfrak{F}$ be a formation and let G be a solvable group. A subgroup F of G is called an **$\mathfrak{F}$-projector** if

(i) $F \in \mathfrak{F}$ and

(ii) if $F \leq H \leq G$ and $N \lhd H$ such that $H/N \in \mathfrak{F}$, then $H = FN$.

Lemma 1.86. *Let F be an $\mathfrak{F}$-projector of a group G.*

(i) *If $F \leq H$ for some subgroup H of G, then F is an $\mathfrak{F}$-projector of H.*

(ii) *If $N \trianglelefteq G$, then FN/N is an $\mathfrak{F}$-projector of G/N.*

(iii) *Let $g \in G$. Then F^g is also an $\mathfrak{F}$-projector of G.*

Beweis. For (i), it is clear that $F \in \mathfrak{F}$. Now let K and N be subgroups of H such that N is normal in K, $F \leq K \leq H$, and $K/N \in \mathfrak{F}$. Since $K \leq G$, $K = FN$ by definition.

To prove (ii), first note that since $F \in \mathfrak{F}$ and $NF/N \cong F/N \cap F$, it follows that $NF/N \in \mathfrak{F}$. Now let H/N and L/N be subgroups of G/N such that $NF/N \leq H/N \leq G/N$ with $L/N \lhd H/N$ and $(H/N)/(L/N) \in \mathfrak{F}$. Since $(H/N)/(L/N) \cong H/L$, the assumption implies that $F \leq H \leq G$ with $L \lhd H$ and $H/L \in \mathfrak{F}$. Since F is an $\mathfrak{F}$-projector for G, $H = FL$ and $H/N = FLN/N$. Thus, $H/N = NF/N \cdot L/N$ and NF/N is an $\mathfrak{F}$-projector for G/N.

For (iii), let H and N be subgroups of G such that $F^g \leq H \leq G$, where $N \lhd H$ and $H/N \in \mathfrak{F}$. First note that $F \leq H^{g^{-1}} \leq G$ and that $N^{g^{-1}} \lhd H^{g^{-1}}$. Since $H^{g^{-1}}/N^{g^{-1}} \cong H/N$ and F is an $\mathfrak{F}$-projector for G, it follows that $H^{g^{-1}} = FN^{g^{-1}}$. Thus, $H = F^g N$ and F^g is an $\mathfrak{F}$-projector for G. □

Lemma 1.87. *Let G be a group, let N be a normal subgroup of G, and let F^*/N be an $\mathfrak{F}$-projector of G/N. If F is an $\mathfrak{F}$-projector of F^*, then F is an $\mathfrak{F}$-projector of G.*

Beweis. Proceed by induction on the order of G. Since F is an $\mathfrak{F}$-projector for F^*, it follows that $F \in \mathfrak{F}$. Let H and L be subgroups of G such that $F \leq H \leq G$ with $L \lhd H$ and $H/L \in \mathfrak{F}$.

First note that $F^*/N \in \mathfrak{F}$ since F^*/N is an $\mathfrak{F}$-projector for G/N. Given that F is an $\mathfrak{F}$-projector for F^* and $F \leq F^*$ yields $F^* = NF$. Second, consider the subgroup $H \cap F^*/H \cap N$ of $H/H \cap N$. It is easy to see that

$$H \cap F^*/H \cap N = H \cap F^*/H \cap F^* \cap N \cong (H \cap F^*)N/N.$$

Applying the modular identity (Lemma 1.6) results in

$$(H \cap F^*)N/N = NH \cap F^*/N = NH \cap NF/N = N(NH \cap F)/N = NF/N = F^*/N,$$

which shows $H \cap F^*/H \cap N \in \mathfrak{F}$.

Now let $K/H \cap N$ and $M/H \cap N$ be subgroups of $H/H \cap N$ such that $H \cap F^*/H \cap N \leq K/H \cap N \leq H/H \cap N$, where $M/H \cap N \lhd K/H \cap N$ and $(K/H \cap N)/(M/H \cap N) \in \mathfrak{F}$. Since $(K/H \cap N)/(M/H \cap N) \cong K/M$, it follows that $K/M \in \mathfrak{F}$. In addition, $F \leq H$ and $F \leq NF = F^*$ implies $F \leq K$. Therefore, $NF/N \leq NK/N \leq NH/N$ or $F^*/N \leq NK/N \leq NH/N$ with $MN/N \lhd NK/N$. Furthermore,

$$(NK/N)/(MN/N) \cong NK/MN = K(MN)/MN \cong K/K \cap MN$$

with

$$M \leq K \cap NM = M(K \cap N) \leq M(H \cap N) = M.$$

Thus, $(NK/N)/(MN/N) \cong K/M$ and $(NK/N)/(MN/N) \in \mathfrak{F}$. Given that F^* is an $\mathfrak{F}$-projector of G/N, it now follows that $NK/N = F^*/N \cdot MN/N$ or $NK = F^*M$. This implies

$$M(H \cap F^*) = MF^* \cap H = NK \cap H = K(N \cap H) = K$$

or that $K/H \cap N = H \cap F^*/H \cap N \cdot M/H \cap N$. Consequently, $H \cap F^*/H \cap N$ is an $\mathfrak{F}$-projector of $H/H \cap N$.

Given that $F \leq H \cap F^*$, Lemma 1.86 implies that F is an $\mathfrak{F}$-projector of $H \cap F^*$. Assume now that $H < G$. Since $H \cap N \lhd H$ with $H \cap F^*/H \cap N$ an $\mathfrak{F}$-projector of $H/H \cap N$, it follows by induction that F is an $\mathfrak{F}$-projector of H. The assumption that $H/L \in \mathfrak{F}$ results in $H = LF$ and that F is an $\mathfrak{F}$-projector of G.

Now suppose $H = G$. Recall that F^*/N is an $\mathfrak{F}$-projector of G/N and that $H/L \in \mathfrak{F}$. Consider the normal subgroup NL of G. Then $NL/L \lhd G/L$ with

$$(G/L)/(NL/L) = (H/L)/(NL/L).$$

The fact that $NL/L \lhd H/L$ with $H/L \in \mathfrak{F}$ implies $(H/L)/(NL/L) \in \mathfrak{F}$ or that $(G/L)/(NL/L) \in \mathfrak{F}$. This results in $G/L = F^*/N \cdot NL/N$. The fact that $G = F^*(NL) = F^*L$ follows immediately.

Given that $F^*/F^* \cap L \cong F^*L/L = G/L$, it follows that $F^*/F^* \cap L \in \mathfrak{F}$ (recall that $G/L = H/L$ with $H/L \in \mathfrak{F}$). Furthermore, since F is an $\mathfrak{F}$-projector of F^*, it follows that $F^* = F(F^* \cap L)$. This means

$$H = G = F^*L = F(F^* \cap L)L = FL$$

and that F is an $\mathfrak{F}$-projector of G. □

Gaschütz [44] proved the following result concerning $\mathfrak{F}$-projectors.

Theorem 1.88. *If $\mathfrak{F}$ is a saturated formation and G a solvable group, then G contains an $\mathfrak{F}$-projector and all $\mathfrak{F}$-projectors are conjugate.*

Beweis. If $G \in \mathfrak{F}$, then the result follows with G as the $\mathfrak{F}$-projector. Suppose $G \notin \mathfrak{F}$, and let N be a minimal normal subgroup of G.

Case 1: $G/N \in \mathfrak{F}$. Given that $G \notin \mathfrak{F}$, Theorem 1.84 implies that N has a complement in G and that all complements are conjugate. Let F be a complement of N in G. Now let H and L be subgroups of G such that $F \leq H \leq G$ with $L \lhd H$ and $H/L \in \mathfrak{F}$. Given that $G = NF$, a simple application of the modular identity (Lemma 1.6) results in $H = H \cap G = H \cap NF = F(H \cap N)$. Since N is abelian, $H \cap N \lhd G$. This means $H \cap N = \{1\}$ or $H \cap N = N$. If $H \cap N = \{1\}$, then $H = F$. If $H \cap N = N$, then $H = G$.

First consider the possibility that $H = F$. In this case, $FL = HL = H$ and F is an $\mathfrak{F}$-projector for G. If $H = G$, then $G/L \in \mathfrak{F}$ as $G/L = H/L$. Now if $N \cap L = \{1\}$, then $G/N \cap L = G$ and $G \in \mathfrak{F}$, a contradiction. Thus, $N \leq L$ and $H = G = NF = LF$, which implies again that F is an $\mathfrak{F}$-projector for G.

Case 2: $G/N \notin \mathfrak{F}$. Proceed by induction on the order of G. Since G/N is solvable, it contains an $\mathfrak{F}$-projector F^*/N. But since $G/N \notin \mathfrak{F}$, it must be that $F^* < G$, where F^* is solvable. Thus, again by induction, F^* has an $\mathfrak{F}$-projector D. By Lemma 1.87, D is an $\mathfrak{F}$-projector for G.

Now let F_1 and F_2 be two $\mathfrak{F}$-projectors for G. Proceed by induction on the order of G. By Lemma 1.86, F_1N/N and F_2N/N are $\mathfrak{F}$-projectors for G/N. Thus, by induction, F_1N/N and F_2N/N are conjugate in G/N. This implies by Lemma 1.14 that there is an element $g \in G$ such that $(F_1N)^g = F_2N$. This means $F_1^g \leq F_2N$. By Lemma 1.86 it follows that F_1^g and F_2 are $\mathfrak{F}$-projectors of F_2N. If $F_2N = G$, then $G/N = F_2N/N$, where $F_2N/N \in \mathfrak{F}$, a contradiction. This results in $F_2N < G$. Thus, by induction, F_1^g and F_2 are conjugate in F_2N, which implies F_1 and F_2 are conjugate in G. □

Recall the concept of a system normalizer that was introduced in Section 1.3. Each system normalizer of a solvable group G is nilpotent, the system normalizers of G form a conjugacy class, and the index of a system normalizer is equal to the number of distinct Sylow systems. In addition, a system normalizer of a solvable group will cover each central chief factor and avoid each eccentric chief factor (Theorem 1.39). In particular, as was shown by Hall [56], if S is a system normalizer of a solvable group G, then $G = L_1S$, where L_1 is the first subgroup in the lower nilpotent series for G.

System normalizers do not need to be self-normalizing. Carter [22] investigated solvable groups whose system normalizers were self-normalizing. A subgroup H of a group G to be **abnormal** if $g \in \langle H, g^{-1}Hg \rangle$ for all $g \in G$. Carter [22] proved that if a system normalizer is self-normalizing, then it is abnormal. He also showed that if the system normalizers of a solvable group G are their own normalizers, then a subgroup H of G is abnormal if and only it contains a system normalizer of G. Carter [23] then proved that if G is a solvable group, it contains nilpotent self-normalizing subgroups. These subgroups were eventually called **Carter subgroups**. He showed that Carter subgroups are abnormal, form a conjugacy class, and that each contains a system normalizer of the group. In certain instances, for example in the class of supersolvable groups, the Carter subgroups coincide with the system normalizers. If $\mathfrak{F} = \mathfrak{N}$ the formation of nilpotent groups, which is a saturated formation, then the $\mathfrak{N}$-projectors are precisely the self-normalizing subgroups or Carter subgroups of G.

In a solvable group G, each chief factor is an elementary abelian p-group for some prime p. Along with the cover and avoidance, the structure of the chief factors is important in determining the structure of G. Recall that for any chief factor H/K of G that $G/C_G(H/K) \cong Aut_G(H/K)$.

Definition 1.89. Associate with each prime p a formation $\mathfrak{F}_p$ with the possibility that $\mathfrak{F}_p$ could be empty. A formation $\mathfrak{F}$ is **locally defined** by $\{\mathfrak{F}_p\}$ or a **locally defined formation** provided $G \in \mathfrak{F}$ if and only if for each prime p dividing the order of G, each p-chief factor H/K of G has $Aut_G(H/K) \in \mathfrak{F}_p$.

Lemma 1.90. *Let $\{\mathfrak{F}_p\}$ be a system of formations that locally defines a formation $\mathfrak{F}$. Then $\{\mathfrak{F}_p \cap \mathfrak{F}\}$ is a systems of formations that locally defines $\mathfrak{F}$ with $\mathfrak{F}_p \cap \mathfrak{F} \subseteq \mathfrak{F}$ for each prime p.*

Beweis. Let G be a group such that $G \in \mathfrak{F}$. Then $Aut_G(H/K) \in \mathfrak{F}_p$ for each p-chief factor H/K of G. Given that $Aut_G(H/K) \cong G/C_G(H/K)$, it also follows that $Aut_G(H/K) \in \mathfrak{F}$. Thus, $Aut_G(H/K) \in \mathfrak{F}_p \cap \mathfrak{F}$. The fact that $\mathfrak{F}_p \cap \mathfrak{F} \subseteq \mathfrak{F}$ follows trivially. □

The next result, which is central to saturated formations, is presented without proof.

Theorem 1.91. *For each saturated formation $\mathfrak{F}$, there is a system of formations $\{\mathfrak{F}_p\}$ that locally defines $\mathfrak{F}$, where $\mathfrak{F}_p \subseteq \mathfrak{F}$.*

Definition 1.92. Given a formation $\mathfrak{F}$, the **$\mathfrak{F}$-residual** of a group G, denoted by $G_{\mathfrak{F}}$, is $\bigcap\{N \mid N \trianglelefteq G \text{ and } G/N \in \mathfrak{F}\}$.

Simply put, $G_{\mathfrak{F}}$ is the smallest normal subgroup of a group G such that $G/G_{\mathfrak{F}} \in \mathfrak{F}$. For the formation $\mathfrak{A}$ of abelian groups, the residual $G_{\mathfrak{A}} = G'$. For the saturated formation $\mathfrak{N}$ of nilpotent groups, the residual $G_{\mathfrak{N}} = L_1(G) = D_\infty(G)$.

Lemma 1.93. *Let $\mathfrak{F}$ be a formation, and let $G_{\mathfrak{F}}$ be the $\mathfrak{F}$-residual and F an $\mathfrak{F}$-projector of a group G.*
(i) *If $N \lhd G$, then $(G/N)_{\mathfrak{F}} = G_{\mathfrak{F}}N/N$.*
(ii) *If $H \leq G$ such that $F \leq H \leq G$, then $H_{\mathfrak{F}} \leq G_{\mathfrak{F}}$.*

Beweis. For (i), let K be a normal subgroup of G such that $G/K \in \mathfrak{F}$. First note that $KN/K \trianglelefteq G/K$ with $(G/K)/(KN/K) \cong G/KN$, which implies $G/KN \in \mathfrak{F}$. Now $KN/N \trianglelefteq G/N$ with $(G/N)/(KN/N) \cong G/KN$, which shows that $(G/N)/(KN/N) \in \mathfrak{F}$. This implies that $(G/N)_{\mathfrak{F}} \leq G_{\mathfrak{F}}N/N$.

Now let H/N be a normal subgroup of G/N such that $(G/N)/(H/N) \in \mathfrak{F}$. The fact that $(G/N)/(H/N) \cong G/H$ implies that $G_{\mathfrak{F}}N/N \leq (G/N)_{\mathfrak{F}}$.

To prove (ii), first assume that $G_{\mathfrak{F}} = \{1\}$. In this case, G is the $\mathfrak{F}$-projector for G. This implies $H = G$ and the result follows.

Suppose $G_{\mathfrak{F}} \neq \{1\}$. Given that $F \leq G$ with $G_{\mathfrak{F}} \trianglelefteq G$ and $G/G_{\mathfrak{F}} \in \mathfrak{F}$, it follows that $G = G_{\mathfrak{F}}F$. Since $F \leq H$, $G = G_{\mathfrak{F}}H$ and $G/G_{\mathfrak{F}} = G_{\mathfrak{F}}H/G_{\mathfrak{F}} = H/H \cap G_{\mathfrak{F}}$. This means $H/H \cap G_{\mathfrak{F}} \in \mathfrak{F}$ and that $H_{\mathfrak{F}} \leq G_{\mathfrak{F}}$. □

Definition 1.94. Let $\mathfrak{F}$ be a saturated formation locally defined by $\{\mathfrak{F}_p\}$ such that $\mathfrak{F}_p \subseteq \mathfrak{F}$ for every prime p. Let Σ be a Sylow system for a group G. An **$\mathfrak{F}$-normalizer** of G associated with Σ is a subgroup of the form

$$\bigcap_p \mathcal{N}_G(G_{\mathfrak{F}_p} \cap H_{p'}),$$

where $H_{p'}$ is the Hall p'-subgroup in Σ.

Carter and Hawkes [26] introduced the concept of a $\mathfrak{F}$-normalizer. If $\mathfrak{F} = \mathfrak{N}$, with $\mathfrak{N}$ the formation of nilpotent groups, then the $\mathfrak{F}$-normalizers coincide with the system normalizers for a solvable group. Furthermore, if $\mathfrak{F}$ is a saturated formation that contains the class of nilpotent groups, the $\mathfrak{F}$-projectors of a group coincide with the $\mathfrak{F}$-normalizers of the group. In addition, given that any two system normalizers are conjugate, it follows that each pair of $\mathfrak{F}$-normalizer are also conjugate.

Definition 1.95. Let formation $\mathfrak{F}$ be locally defined by $\{\mathfrak{F}_p\}$. A p-chief factor H/K of G is $\mathfrak{F}$-**central** if $Aut_G(H/K) \in \mathfrak{F}_p$. If H/K is not $\mathfrak{F}$-central it is $\mathfrak{F}$-**eccentric**.

The next result lays the foundation for a number of normal subgroup complementation results stated in Chapter 7. However, its proof is omitted as it requires a number of ideas that, while important, get a bit off-topic.

Lemma 1.96. *Let $\mathfrak{F}$ be a formation locally defined by $\{\mathfrak{F}_p\}$ such that $\mathfrak{F}_p \subseteq \mathfrak{F}$ for every prime p. Let Σ be a Sylow system for a group G, and let $A_p = G_{\mathfrak{F}_p} \cap H_{p'}$, where $H_{p'}$ is the Hall p'-subgroup in Σ. Then $\mathcal{N}_G(A_p)$ covers every $\mathfrak{F}$-central p-chief factor and avoids every $\mathfrak{F}$-eccentric p-chief factor.*

The importance of Lemma 1.96 is demonstrated in the next result.

Theorem 1.97. *Let A be an $\mathfrak{F}$-normalizer of a group G. Then A covers each $\mathfrak{F}$-central chief factor and avoids each $\mathfrak{F}$-eccentric chief factor.*

Beweis. Let Σ be a Sylow system for G. Then $A = \bigcap_p \mathcal{N}_G(G_{\mathfrak{F}_p} \cap H_{p'})$, where $H_{p'}$ is the Hall p'-subgroup in Σ. Let $A_p = G_{\mathfrak{F}_p} \cap H_{p'}$. By its definition, A will avoid all p-chief factors of G that $\mathcal{N}_G(A_p)$ avoids. By Lemma 1.96, A avoids all $\mathfrak{F}$-eccentric chief factors.

Now note that $A_p \lhd H_{p'}$ for each prime p. This implies $H_{p'} \leq \mathcal{N}_G(A_p)$ and $[G : \mathcal{N}_G(A_p)]$ is a power of p. Let

$$1 = N_0 \lhd N_1 \lhd \cdots \lhd N_t = G$$

be a chief series for G, and let N_{i_j}/N_{i_j-1} be the $\mathfrak{F}$-central p-chief factors in this series where $|N_{i_j}/N_{i_j-1}| = p^{\alpha_{i_j}}$ for $1 \leq j \leq n$. Given that $\mathcal{N}_G(A_p)$ covers these chief factors (Lemma 1.96), $N_{i_j} \leq N_{i_j-1}\mathcal{N}_G(A_p)$ and $p^{\alpha_{i_j}}$ divides $|\mathcal{N}_G(A_p)|$. Now let q be a prime divisor of the order of G such that $q \neq p$. Given that $[G : \mathcal{N}_G(A_q)]$ is a power of q, $p^{\alpha_{i_j}}$ divides $|\mathcal{N}_G(A_p)|$ and $p^{\alpha_{i_1}+\cdots+\alpha_{i_n}}$ divides $|A|$.

Let $\alpha = \alpha_{i_1} + \cdots + \alpha_{i_n}$, and suppose p^t divides $|A|$, where $t > \alpha$. Then for some $\mathfrak{F}$-eccentric p-chief factor N_l/N_{l-1}, $\mathcal{N}_G(A_p) \cap N_l \nleq N_{l-1}$, which is a contradiction by Lemma 1.96. Thus, $|A|$ is the product of the orders of the $\mathfrak{F}$-central chief factors of any chief series for G.

Consider again the chief series

$$1 = N_0 \lhd N_1 \lhd \cdots \lhd N_t = G$$

for G. It follows that

$$|A| = |N_0 \cap A||N_1 \cap A/N_0 \cap A| \cdots |N_t \cap A/N_{t-1} \cap A|.$$

Consider an arbitrary chief factor N_i/N_{i-1}. If N_i/N_{i-1} is $\mathfrak{F}$-eccentric, then A avoids N_i/N_{i-1}, which implies $N_i \cap A \leq N_{i-1}$. This means $N_i \cap A/N_{i-1} \cap A = \{1_{N_{i-1} \cap A}\}$ or that $|N_i \cap A/N_{i-1} \cap A| = 1$.

If N_i/N_{i-1} is $\mathfrak{F}$-central, then

$$|N_i \cap A/N_{i-1} \cap A| = |N_i \cap A/(N_i \cap A) \cap N_{i-1}| = |N_{i-1}(N_i \cap A)/N_{i-1}| \leq |N_i/N_{i-1}|.$$

However, the order of A is the product of the orders of the $\mathfrak{F}$-central chief factors, which implies that $|N_{i-1}(N_i \cap A)/N_{i-1}| = |N_i/N_{i-1}|$ or that $N_{i-1}(N_i \cap A) = N_i$. Thus, A covers N_i/N_{i-1}. □

Lemma 1.98. *If N is a normal subgroup of a group G and A is an $\mathfrak{F}$-normalizer of G, then AN/N is an $\mathfrak{F}$-normalizer of G/N.*

Beweis. Since A is an $\mathfrak{F}$-normalizer of G, there is a Sylow system Σ for G such that $A = \bigcap_p \mathcal{N}_G(G_{\mathfrak{F}_p} \cap H_{p'})$, where $H_{p'}$ is the Hall p'-subgroup in Σ. It follows by Lemma 1.93 that $(G/N)_{\mathfrak{F}_p} = G_{\mathfrak{F}_p}N/N$. Now $H_{p'}N/N$ is a Hall p'-subgroup of G/N, where

$$\begin{aligned}(G/N)_{\mathfrak{F}_p} \cap H_{p'}N/N &= G_{\mathfrak{F}_p}N/N \cap H_{p'}N/N \\ &= G_{\mathfrak{F}_p}N \cap H_{p'}N/N.\end{aligned}$$

Since A normalizes $G_{\mathfrak{F}_p}N$ and $H_{p'}N$, it follows that

$$AN/N \leq \bigcap_p \mathcal{N}_{G/N}((G/N)_{\mathfrak{F}_p} \cap H_{p'}N/N).$$

Now let

$$1 = N_0 \lhd N_1 \lhd \cdots \lhd N_i = N \lhd N_{i+1} \lhd \cdots \lhd N_t = G$$

be a chief series for G containing N. Then

$$N/N = N_i/N \lhd N_{i+1}/N \lhd \cdots \lhd N_t/N = G/N$$

is a chief series for G/N with $(N_j/N)/(N_{j-1}/N) \cong N_j/N_{j-1}$ for $i+1 \le j \le t$. Given that

$$\begin{aligned}
|NA/N| &= |N_i/N \cap NA/N||(N_{i+1}/N \cap NA/N)/(N_i/N \cap NA/N)| \\
&\qquad \cdots |(N_t/N \cap NA/N)/(N_{t-1}/N \cap NA/N)| \\
&= |N_i \cap NA/N||(N_{i+1} \cap NA/N)/(N_i \cap NA/N)| \\
&\qquad \cdots |(N_t \cap NA/N)/(N_{t-1} \cap NA/N)| \\
&= |N(N_i \cap A)/N||(N(N_{i+1} \cap A)/N)/(N(N_i \cap A)/N)| \\
&\qquad \cdots |(N(N_t \cap A)/N)/(N(N_{t-1} \cap A)/N)| \\
&= |N_i \cap A/N \cap N_i \cap A||(N_{i+1} \cap A/N \cap N_{i+1} \cap A)/(N_i \cap A)/N_i \cap N \cap A)| \\
&\qquad \cdots |(N_t \cap A/N \cap N_t \cap A)/(N_{t-1} \cap A/N \cap N_{t-1} \cap A)| \\
&= |N_i \cap A/N_i \cap A||(N_{i+1} \cap A/N \cap A)/(N_i \cap A)/N \cap A)| \\
&\qquad \cdots |(N_t \cap A/N \cap A)/(N_{t-1} \cap A/N \cap A)| \\
&= |N_i \cap A/N_i \cap A||N_{i+1} \cap A/N_i \cap A| \cdots |N_t \cap A/N_{t-1} \cap A|,
\end{aligned}$$

It follows from Theorem 1.97 and its proof that $|NA/N|$ is the product of the orders of the $\mathfrak{F}$-central chief factors of G/N. This implies that NA/N is an $\mathfrak{F}$-normalizer of G/N. □

Theorem 1.99. *Let G be a solvable group. Then each $\mathfrak{F}$-projector of G contains an $\mathfrak{F}$-normalizer of G.*

Beweis. Proceed by induction on the order of G. Let F be an $\mathfrak{F}$-projector of G, and let A be an $\mathfrak{F}$-normalizer of G. Note that there exists a Sylow system Σ that reduces into F (see the comment after Lemma 1.30 in Section 1.3). If $F = G$, the result follows. Assume that $F \neq G$, which means $G \notin \mathfrak{F}$. Let N be a minimal normal subgroup of G. Then by Lemma 1.86, FN/N is an $\mathfrak{F}$-projector of G/N, and by Lemma 1.98, AN/N is an $\mathfrak{F}$-normalizer of G/N. By induction, FN/N contains an $\mathfrak{F}$-normalizer of G/N. But given that all $\mathfrak{F}$-normalizers are conjugate, it can be assumed that $AN/N \le FN/N$ or that $A \le FN$.

First consider the possibility that $FN \neq G$. Let $H_{p'}$ be the Hall p'-subgroup in Σ. Then $L_{p'} = H_{p'} \cap FN$ is a Hall p'-subgroup of FN. Recall that F is an $\mathfrak{F}$-projector of G. Since $G/G_{\mathfrak{F}_p} \in \mathfrak{F}_p$ and $\mathfrak{F}_p \subseteq \mathfrak{F}$ (see Definition 1.94), it follows that $G/G_{\mathfrak{F}_p} \in \mathfrak{F}$ and that $G = FG_{\mathfrak{F}_p}$. Consequently,

$$FN/FN \cap G_{\mathfrak{F}_p} \cong FNG_{\mathfrak{F}_p}/G_{\mathfrak{F}_p} = G/G_{\mathfrak{F}_p}$$

and $FN/FN \cap G_{\mathfrak{F}_p} \in \mathfrak{F}_p$. This means $FN_{\mathfrak{F}_p} \leq FN \cap G_{\mathfrak{F}_p}$ or that $FN_{\mathfrak{F}_p} \leq G_{\mathfrak{F}_p}$. Thus, for any prime p, it follows that

$$\begin{aligned} G_{\mathfrak{F}_p} \cap H_{p'} \cap FN_{\mathfrak{F}_p} &= H_{p'} \cap FN_{\mathfrak{F}_p} \\ &= L_{p'} \cap FN \cap FN_{\mathfrak{F}_p} \\ &= L_{p'} \cap FN_{\mathfrak{F}_p}. \end{aligned}$$

Given that F is an $\mathfrak{F}$-projector of FN (Lemma 1.86) and $FN < G$, it follows by induction that $\bigcap_p \mathcal{N}_{FN}(L_{p'} \cap FN_{\mathfrak{F}_p}) \leq F$.

Let $a \in A$. Then the fact that $(H_{p'} \cap G_{\mathfrak{F}_p})^a = H_{p'} \cap G_{\mathfrak{F}_p}$ implies $(H_{p'} \cap G_{\mathfrak{F}_p} \cap FN_{\mathfrak{F}_p})^a = H_{p'} \cap G_{\mathfrak{F}_p} \cap FN_{\mathfrak{F}_p}$ or that $(L_{p'} \cap FN_{\mathfrak{F}_p})^a = L_{p'} \cap FN_{\mathfrak{F}_p}$. This implies $A \leq F$.

Now suppose that $G = FN$. Since G is solvable, N is an elementary abelian q-group for some prime q. Since $G = FN$ with N abelian, $F \cap N \lhd G$, which implies that $F \cap N = \{1\}$ and that F is maximal in G. Recall that $G \notin \mathfrak{F}$. However, since $G/N \cong F$ with $F \in \mathfrak{F}$ and N minimal normal in G, this means $G_{\mathfrak{F}} = N$. Suppose K is another minimal normal subgroup of G. If $FK < G$, then by the work already completed, it can be shown that $A \leq F$. If $FK = G$, then by an argument identical to the one just given, $G_{\mathfrak{F}} = K$ or $N = K$.

Assume that N is the unique minimal normal subgroup of G. Let $Q/N = O_{q'}(G/N)$, and let $J_{q'}$ be a Hall q'-subgroup of Q. Thus, $J_{q'} \leq F$, which implies $Q \cap F = J_{q'}$. Given that $Q \lhd G$ implies $Q \cap F \lhd F$, it follows that $J_{q'} \lhd F$. Since N is the unique minimal normal subgroup of G of order a power of q, $J_{q'}$ is not normal in G. Given that $G = FN$, it follows that $\mathcal{N}_G(J_{q'}) = F$.

Note that $[G : F]$ is a power of q. Let $H_{q'}$ be a Hall q'-subgroup in Σ. Then $H_{q'} \cap F$ is a Hall q'-subgroup of F or $H_{q'} \leq F$. This means $H_{q'} \cap G_{\mathfrak{F}_p} \leq F$.

Let $R/N = O_{q'q}(G/N)$, recalling that $G/N \in \mathfrak{F}$. Since $G/R \cong (G/N)/(R/N)$ with $(G/N)/(R/N) \in \mathfrak{F}$, it follows that $G/R \in \mathfrak{F}$. Let $(H/N)/(K/N)$ be a q-chief factor of G/N. Then $Aut_G((H/N)/(K/N)) = (G/N)/C_{G/N}((H/N)/(K/N)) \in \mathfrak{F}_p$, which implies that $(G/N)/\cap C_{G/N}((H/N)/(K/N)) \in \mathfrak{F}_p$ over all q-chief factors of G/N. By Theorem 3.2 on page 216 in [106], $O_{q'q}(G/N) = \bigcap C_{G/N}((H/N)/(K/N))$, which results $(G/N)/(R/N) \in \mathfrak{F}_p$ or that $G/R \in \mathfrak{F}_p$. Consequently, $G_{\mathfrak{F}_p} \leq R$. It then follows that $H_{q'} \bigcap G_{\mathfrak{F}_p} \leq Q$.

Recall that $H_{q'} \cap G_{\mathfrak{F}_p} \leq F$. This means that $H_{q'} \cap G_{\mathfrak{F}_p} \leq J_{q'}$ and that

$$\begin{aligned} H_{q'} \cap G_{\mathfrak{F}_p} &= J_{q'} \cap H_{q'} \cap G_{\mathfrak{F}_p} \\ &= J_{q'} \cap G_{\mathfrak{F}_p}. \end{aligned}$$

Since $J_{q'} \lhd F$ and $G_{\mathfrak{F}_p} \lhd G$, it follows that $H_{q'} \cap G_{\mathfrak{F}_p} = J_{q'} \cap G_{\mathfrak{F}_p}$ is normal in F. Since F is maximal in G, $\mathcal{N}_G(H_{q'} \cap G_{\mathfrak{F}_p}) = F$ or $\mathcal{N}_G(H_{q'} \cap G_{\mathfrak{F}_p}) = G$. If $\mathcal{N}_G(H_{q'} \cap G_{\mathfrak{F}_p}) = G$, then $H_{q'} \cap G_{\mathfrak{F}_p}$ is normal in G with $N \leq H_{q'} \cap G_{\mathfrak{F}_p}$. This is a contradiction. Thus, $\mathcal{N}_G(H_{q'} \cap G_{\mathfrak{F}_p}) = F$, which implies $A \leq F$. □

Since any two $\mathfrak{F}$-normalizers of a solvable group are conjugate, the next result immediately follows.

Corollary 1.100. *If G is a solvable group, then any $\mathfrak{F}$-normalizer of G is contained in an $\mathfrak{F}$-projector of G.*

2 The Schur-Zassenhaus theorem: A bit of history and motivation

This chapter documents the development of the Schur-Zassenhaus theorem, one of the most well-known results in group theory, from Schur's initial 1902 result to full version by which it is known today. Because Schur's 1902 result was the seed that grew into the field of normal subgroup complementation, more time will be spent in this chapter than in the other chapters on the historical development of the theorems presented in the chapter. In addition, several generalizations of the Schur-Zassenhaus theorem will also be presented.

Before getting to Schur's 1902 inspirational result, the stage needs to be set by Hölder and Frobenius. In the late 1800s and early 1900s, one main thrust in finite group theory was the study of group extensions. This area got its start in an 1893 paper by Hölder [62] where he classified the groups of order p^3, pq^2, pqr, and p^4 for distinct primes p, q, and r. He stated the problem more succinctly in an 1895 paper where he was concerned with the composition series of a finite group.

> One must solve the following problem several times: To form a group Δ which contains a given normal subgroup Γ, such that Δ/Γ corresponds to a given group (page 322 in [63], author's translation).

This central question in extension theory was stated in a more focused way in 1926 by Schreier. As he discussed on page 12 of his paper [94], the goal, when given two groups N and H, is to determine those groups G which have a normal subgroup M with $M \cong N$ and $G/M \cong H$. Schreier [93, 94] wrote a series of two papers on the subject. One component of extension theory was to find a complement to M in G which would be, by definition, isomorphic to the H.

One way in which Frobenius helped set the stage for Schur's result is the simple fact that he was Schur's thesis advisor. However, more important are his contributions to finite group theory that motivated Schur's work. At the end of the 19th and early 20th centuries, Frobenius had already made many significant contributions to group theory. In particular, he studied a specific class of permutation groups which were named after him.

Definition 2.1. A transitive permutation group G where only the identity element fixes more than one letter and the subgroup fixing one letter is nontrivial is called a **Frobenius group**.

In 1901, Frobenius proved (V in [40]) that in a Frobenius group, the collection of elements that fix no letters, together with the identity element, forms a normal subgroup K of G. This subgroup K is called the **Frobenius kernel** of G and Thompson

DOI 10.1515/9783110480214-002

[105] proved that K is nilpotent. Frobenius groups also have a nice decomposition, as shown in the following theorem.

Theorem 2.2. [Frobenius] *Let G be a Frobenius group with Frobenius kernel K, and let H be a subgroup of G fixing one letter. Then, $G = [K]H$.*

Theorem 2.2 and its proof, which is omitted here, motivates the following two alternate definitions of a Frobenius group.

Definition 2.3. A group G is a **Frobenius group** if there exists a proper and nontrivial subgroup H of G, such that $H \cap g^{-1}Hg = \{1\}$ for all $g \in G \backslash H$.

Definition 2.4. A group G is a **Frobenius group** if there exists a nontrivial and proper normal subgroup K of G, such that if $x \in K \backslash \{1\}$, then $C_G(x) \subset K$. In this case, K is called the **Frobenius kernel**.

Frobenius also more directly motivated Schur's 1902 result. In 1901, Frobenius proved the following theorem.

Theorem 2.5. [I in Frobenius [40]] *If a group H of order $h = gn$ contains a subgroup G of order g, and if every two elements from G, which are conjugate in H, are also conjugate in G, then the nth power of every linear character of G is a linear character of H.*

If r is the order and m the index of the commutator subgroup R of G, and if m and n are relatively prime, then the elements of H, whose orders divide n, together with the commutator subgroup of H, generate a characteristic subgroup S of H, whose order s is divisible by r and n and whose index is divisible by m. If g and n are relatively prime, then $s = rn = \frac{h}{m}$, and the commutative group H/S is isomorphic to the group G/R (page 1219 in [40], author's translation).

The main goal of Schur's 1902 paper was to give a new proof of this result that did not rely on character theory. This may explain why Schur stated Frobenius's result (Theorem 2.5) as follows:

Theorem 2.6. [I from Frobenius [40] as stated by Schur [95]] *If a group H of order $h = gn$ contains a subgroup G of order g, if every two elements from G, which are conjugate in H, are also conjugate in G, if r is the order and m the index of the commutator subgroup R of G, and if m and n are relatively prime, then the elements of H, whose orders divide n, together with the commutator subgroup of H, generate a characteristic subgroup S of H, whose order s is divisible by r and n and whose index is divisible by m. If g and n are relatively prime, then $s = rn = \frac{h}{m}$, and the commutative group H/S is isomorphic to the group G/R* (page 1013 in [95], author's translation).

Schur then used the techniques he used to prove Frobenius's result to prove the following theorem.

Theorem 2.7. [III in Schur [95]] *If a group G of order nm contains a normal subgroup N of order n, such that N is contained in the center of the G with $(n, m) = 1$, then G is the direct product of N with a subgroup M of order m.*

Given that Schur's result is a corollary to Theorem 2.9 (presented below), its proof is omitted. However, a few comments must be made before proceeding. First, note that since N is contained in the center of the group, it is abelian. While the author can find no evidence in the literature of such a result by Schur, he is often given credit (e.g. Bergström [16] and Gaschütz [41]) for the following stronger result.

Theorem 2.8. *Let N be a normal abelian subgroup of a group G whose order is relatively prime to its index, then N has a complement in G.*

As with Schur's result (Theorem 2.7), Theorem 2.8 is a corollary to Theorem 2.9 and its proof is omitted. A few even give him credit for the following even stronger result (e.g. Dixon [33]).

Theorem 2.9. *Let N be a normal abelian subgroup of a group G whose order is relatively prime to its index, then N has a complement in G and all complements of N in G are conjugate.*

Beweis. First, consider the case that $[G : N]$ is the power of some prime p. Let P be a Sylow p-subgroup of G. By Lemma 1.2, NP is a subgroup of G. Since $|NP| = |N||P|/|N \cap P|$, $G = NP$ and G splits over N. Given that all Sylow p-subgroups are conjugate in G, all complements of N in G are conjugate.

Now assume that $l = [G : N]$ is divisible by two or more primes, with p being one of those primes. Let P be a Sylow p-subgroup of G. By Lemma 1.2, $B = NP = [N]P$ is a proper subgroup of G. Given that $[G : B]$ divides l, Theorem 1.80 implies that G splits over N. Furthermore, given that all complements of N in B are conjugate, Theorem 1.80 also implies that all conjugates of N in G are conjugate. □

The only person the author is familiar with who appears to cite Schur's 1902 result correctly is Suzuki (page 235 of [103]). However, this theme of finding complements to abelian normal subgroups reoccurs many times in the development of this topic.

The next observation is that in Theorem 2.7, the order of N is relatively prime to its index. Originally, this requirement that the order of the normal subgroup be relatively prime to its index is needed so that the factor system would behave in such a way so that a complement for the normal subgroup could be found. A number of results in normal subgroup complementation focus on normal subgroups with this, or a generalization of this, condition which require the order of N to be relatively prime to its index.

The first generalization of Schur's result was by Burnside and appeared in the 1911 second edition [21] of his book *Theory of Groups of Finite Order* [18]. It should be noted that Burnside proved this result as a corollary to a theorem on self-conjugate (normal) subgroups and that his result (Theorem 2.10) was not contained in the first edition of his book [18] published in 1897.

Theorem 2.10. [Corollary 1 on page 327 in Burnside [21]] *If $|G| = p^\alpha m$, where m is not divisible by p, and if every element of G whose order is a power of p commutes with every element whose order is relatively prime to p, then G is the direct product of two groups of orders p^α and m.*

Beweis. Let P be a Sylow p-subgroup of G. It follows from the hypothesis that P is normal in G. If P is abelian, then $P \leq Z(G)$ and the result follows from Schur (Theorem 2.7).

Now consider the case that P is nonabelian and proceed by induction on the order of G. Since $P' \lhd G$, consider the normal subgroup P/P' of G/P' of order p^β. Clearly $\beta < \alpha$ and P/P' satisfies the conditions of Schur's result (Theorem 2.7). Thus, there exists a subgroup H/P' of G/P' having order m satisfying $G/P' = P/P' \times H/P'$. Given that $|H| = p^\beta m < |G|$, there exists a subgroup K of H having order m and satisfying $H = P' \times K$. It follows directly that $G = P \times K$. □

Both Schur's (Theorem 2.7) and Burnside's results (Theorem 2.10) were generalized by Weisner in 1927. Weisner, while not citing Schur's result, does indicate that he is proving a generalization of Burnside's result.

Theorem 2.11. [Main Theorem in Weisner [107]] *A group G of order mn, where m and n are relatively prime, in which every element whose order divides m is commutative with every element whose order divides n, is the direct product of two groups of order m and n.*

Weisner's Proof. Proceed by induction on the order of the group G. Assume that m, $n > 1$.

First, consider the case that there exists an element in G of order m. Let p^α be that largest power of p, for p a prime that divides m. Then, every element whose order is a power of p will commute with every element of order prime to p. Thus, by Burnside's result (Theorem 2.10), G contains a subgroup H of order $\frac{mn}{p^\alpha}$. Since $|H| < |G|$, with p^α dividing $|M|$, there exists a subgroup K of H, such that $|K| = n$.

Next consider the case that there are no elements in G of order m. Then, there exists an element $s \in G$, such that $|s| = m_1$, where $m_1|m$ and $m_1 < m$. The subgroup $\mathcal{N}_G(\langle s\rangle)$ contains all of the Sylow subgroups whose orders divide n. This implies that $|\mathcal{N}_G(\langle s\rangle)| = nm_2$, where $m_2|m$ with $m_2 \geq m_1$. If $m_2 < m$, then $|\mathcal{N}_G(\langle s\rangle)| < |G|$ and $\mathcal{N}_G(\langle s\rangle)$ contains a subgroup K of order n.

Now assume that $m_2 = m$, which implies $\langle s\rangle \lhd G$. Since $|G/\langle s\rangle| = \frac{nm}{m_1}$, let $t\langle s\rangle$ be an element of $G/\langle s\rangle$, such that $|t\langle s\rangle| = n_1$, where $n_1|n$. Then, $t \in G$ with $|t| = n_1 r$, where $r|m$. Given that $(n_1, r) = 1$, there exists two elements t_1 and s_1 in G, such that $t = t_1 s_1$, where $|t_1| = n_1$ and $|s_1| = r$. The calculation $t^{n_1} = (t_1 s_1)^{n_1} = t_1^{n_1} s_1^{n_1} = s_1^{n_1} \in \langle s\rangle$, where $(n_1, |\langle s\rangle|) = 1$, implies that there exists an integer q, such that $s_1 = s^q$. This results in $t = t_1 s^q$.

For all elements $g \in G$, such that $|g\langle s\rangle|$ divides $\frac{m}{m_1}$, then $|g||m$ and $gt_1 = t_1 g$. It follows that

$$g\langle s\rangle t\langle s\rangle = g\langle s\rangle t_1\langle s\rangle = gt_1\langle s\rangle = t_1 g\langle s\rangle = t_1\langle s\rangle g\langle s\rangle = t\langle s\rangle g\langle s\rangle$$

and every element in $G/\langle s\rangle$ whose order divides n commutes with every element whose order divides $\frac{m}{m_1}$. By induction, $G/\langle s\rangle$ contains a subgroup $H/\langle s\rangle$, such that $|H/\langle s\rangle| = n$. Thus, H is a subgroup of G with $|H| = nm_1 < nm$. By induction, there exists a subgroup K of H with $|K| = n$.

In each case, G has a subgroup N of order n. A similar argument results in a subgroup M of G, where $|M| = m$. This implies that $G = N \times M$. □

Next comes the initial form of the Schur-Zassenhaus theorem. It was originally stated in two theorems which first appeared in 1937 in Zassenhaus's book *Lehrbuch der Gruppentheorie* [114]. He gave Schur credit for the first of the two theorems (Theorem 2.12), but did not provide a reference. The author can find no evidence of such a result by Schur in the literature. Zassenhaus may have credited Schur as he used the same proof techniques as Schur did. A few mathematicians took Zassenhaus's lead (e.g. D.G. Higman [59]) and credited Schur with Theorem 2.12, while others (e.g. Huppert [66]) do credit Zassenhaus with Theorem 2.12. It is a vast improvement of Schur's original result (the normal subgroup need not be contained in the center of the group or be abelian).

Theorem 2.12. [Satz IV.7.25 on page 125 in Zassenhaus [114]] *Let N be a normal subgroup of a group G whose order is relatively prime to its index, then N has a complement in G.*

Beweis. Proceed by induction on the order of G. First, it will be shown that N has a proper supplement in G. Start with the case that N is nilpotent. Let P be a Sylow p-subgroup of N, where p divides the order of N. If P is abelian, then by Theorem 2.9, $G = [P]H$ for $H < G$, which implies that $G = NH$. Suppose P is not abelian. Since P is normal in G, $P' \lhd G$ and P/P' is an abelian normal subgroup of G/P' with $([G/P' : P/P'], |P/P'|) = 1$. Thus, by Theorem 2.9, $G/P' = [P/P']K/P'$ for some subgroup K of G. This implies $G = PK = NK$ for $K < G$.

Now consider the case that N is not nilpotent. Then, by Lemma 1.46, there is a Sylow p-subgroup P of N, where p divides the order of N, which is not normal in G. By the Frattini Argument (Lemma 1.7), $G = N\mathcal{N}_G(P)$, with $\mathcal{N}_G(P)$ being a proper subgroup of G.

In all cases, there exists a proper subgroup L of G, such that $G = NL$ with $N \cap L \lhd L$. Since $|NL/L| = |L/(N \cap L)|$, it follows that $([L:N \cap L], |N \cap L|) = 1$. By induction, there is a subgroup M of L, such that $L = [N \cap L]M$. Then, by Lemma 1.11, $G = [N]M$ and G splits over N. □

Zassenhaus also addressed the conjugacy of complements question.

Theorem 2.13. [Satz IV.7.27 on page 126 in Zassenhaus [114]] *Let N be a normal subgroup of a group G whose order is relatively prime to its index, then any two complements are conjugate if one of the following conditions holds:*
(i) *N is abelian.*
(ii) *N is solvable.*
(iii) *G/N is solvable.*

Beweis. Given that item (i) follows from Theorem 2.9, assume that N is not abelian. For items (ii) and (iii), suppose that $G = [N]H = [N]K$ for subgroups H and K of G and proceed by induction on the order of G.

First, consider the case that N is solvable. Then, N' is a proper nontrivial subgroup of N and is normal in G. Thus, N/N' is an abelian normal subgroup of G/N', such that $([G/N' : N/N'], |N/N'|) = 1$. Furthermore, by Lemma 1.11, $G/N' = [N/N']N'H/N' = [N/N']N'K/N'$. Thus, by Theorem 2.9, there is an element $g \in G$, such that $(N'H)^{gN'} = N'K$. This implies $(N'H)^g = N'K$ or that $N'H^g = N'K$. Given that $N' < N$, $B = [N']H^g = [N']K$ is a proper subgroup of G. By induction, there is an element $b \in B$, such that $(H^g)^b = K$ or that $H^{gb} = K$. As a result, H and K are conjugate in G.

Now suppose that G/N is solvable. Since $H \cong G/N \cong K$, there is an isomorphism $\alpha : H \to K$ defined by $\alpha(h)N = kN$ for $h \in H$ and $k \in K$. Given that $N \cap H = N \cap K = \{1\}$ and $G = NH$, for each $h \in H$, $\alpha(h) = k = nh'$ for some $n \in N$ and $h' \in H$.

Since $H \cong G/N$ is solvable, H contains a minimal normal subgroup M, such that M is an elementary abelian p-group for prime p dividing the order of H. If $M = H$, then H and K are Sylow p-subgroups of G and are conjugate in G. Proceed assuming that $M < H$. Then, $B = [N]M = [N]\alpha(M)$ is a proper subgroup of G. By induction, there an element $b \in B$, such that $M = (\alpha(M))^b$. Given that $\alpha(M) \lhd K$, $M = (\alpha(M))^b \lhd K^b$ and both H and K^b are contained in $\mathcal{N}_G(M)$.

Now by the modular identity (Lemma 1.6)

$$\mathcal{N}_G(M) = NH \cap \mathcal{N}_G(M) = H(N \cap \mathcal{N}_G(M)).$$

Since $N \cap \mathcal{N}_G(M) \lhd \mathcal{N}_G(M)$ with $H \cap (N \cap \mathcal{N}_G(M)) = \{1\}$, it follows that $\mathcal{N}_G(M) = [N \cap \mathcal{N}_G(M)]H$ with $([\mathcal{N}_G(M) : N \cap \mathcal{N}_G(M)], |N \cap \mathcal{N}_G(M)|) = 1$. A similar calculation shows that $\mathcal{N}_G(M) = [N \cap \mathcal{N}_G(M)]K^b$ with $([\mathcal{N}_G(M) : N \cap \mathcal{N}_G(M)], |N \cap \mathcal{N}_G(M)|) = 1$.

For the last step, consider $\mathcal{N}_G(M)/M$. By Lemma 1.11,

$$\mathcal{N}_G(M)/M = [(N \cap \mathcal{N}_G(M))M/M]H/M = [(N \cap \mathcal{N}_G(M))M/M]K^b/M,$$

where

$$([\mathcal{N}_G(M)/M : (N \cap \mathcal{N}_G(M))M/M], |(N \cap \mathcal{N}_G(M))M/M|) = 1.$$

By induction, H/M and K^b/M are conjugate in $\mathcal{N}_G(M)/M$. This implies that H and K^b, or that H and K, are conjugate in G. □

Before continuing with the discussion on the development of the Schur-Zassenhaus theorem, three simple applications of Theorem 2.12 and Theorem 2.13 are given. They will be needed later on, but first a definition.

Definition 2.14. Let G be a group and let H be a subgroup of G. The **normal closure of H in G**, denoted by H^G, is the normal subgroup $\langle h^g \mid h \in H \text{ and } g \in G\rangle$ of G.

Equivalently, the normal closure of a subgroup H in group G is the smallest normal subgroup of G containing H.

Now let π be a subset of the set of primes dividing the order of a group G, and let N be a normal Hall π-subgroup of G. If G has a Hall π'-subgroup H, then $G = [N]H$ as $N \cap H = \{1\}$ and $NH \leq G$ (Lemma 1.2) with $|NH| = |N||H|/|N \cap H| = |G|$. This motivates the following lemma.

Lemma 2.15. *Let G be a group with a normal abelian Sylow p-subgroup P, and let H be a complement to P in G. If $H^G = G$, then $\mathcal{N}_G(H) = H$.*

Beweis. Let $P^* = P \cap \mathcal{N}_G(H)$. Since P is abelian and normal in G, P^* is normal in $\mathcal{N}_G(H)$. By the modular identity (Lemma 1.6),

$$\mathcal{N}_G(H) = G \cap \mathcal{N}_G(H) = PH \cap \mathcal{N}_G(H) = (P \cap \mathcal{N}_G(H))H = P^*H,$$

and it follows that $\mathcal{N}_G(H) = P^* \times H$. Given that $G = PH$ with P abelian and P^* in the center of $\mathcal{N}_G(H)$, $P^* \leq Z(G)$.

Suppose p divides the order of G/G'. This implies that $O^p(G) \neq G$. By Lemma 1.27, $O^p(G)$ has a Hall p'-subgroup K. Then, K is a Hall p'-subgroup of G and a complement of P. Thus, by Theorem 2.13, K and H are conjugate in G and $H \leq O^p(G)$. This is a contradiction as $H^G = G$. Consequently, p does not divide the order of G/G' and $P \leq G'$.

This implies $P^* \leq G' \cap Z(G) \cap P$. Given that $G' \cap Z(G) \cap P = \{1\}$ (see IV.2.2 in [66]), $P \cap \mathcal{N}_G(H) = \{1\}$ and $\mathcal{N}_G(H) = H$. □

This leads to the following result, which is a variation of the Frattini Argument (Lemma 1.7).

Lemma 2.16. *If N is a normal subgroup of a group G, and if N has a normal Sylow p-subgroup P, then $G = N\mathcal{N}_G(H)$ for any Hall p'-subgroup H of N.*

Beweis. Let $g \in G$. By Lemma 1.27, H^g is a Hall p'-subgroup of N. Since both H and H^g are complements of P in N, it follows from Theorem 2.13 that H and H^g are conjugate in N. This means there exists an element $n \in N$, such that $H^g = H^n$ or $H^{gn^{-1}} = H$. Thus, $gn^{-1} \in \mathcal{N}_G(H)$ or $gn^{-1} = h$, where $h \in \mathcal{N}_G(H)$. This implies $g = hn$ or that $G = N\mathcal{N}_G(H)$. □

Lemma 2.17. *Let N be a normal subgroup of a group G, such that $G = NH$, where $H \leq G$ and $([H : N \cap H], |N \cap H|) = 1$. Then, G splits over N.*

Beweis. If $N \cap H = \{1\}$, the result follows. Assume otherwise. Given that $N \cap H$ is normal in H and relatively prime to its index, there exists by Theorem 2.12 a subgroup K of H, such that $H = [N \cap H]K$. Thus, by Lemma 1.11, $G = [N]K$ and G splits over N. □

Now Zassenhaus believed that none of the conditions stated in Theorem 2.13 were required for all complements to be conjugate as it had already been conjectured by Burnside that all groups of odd order were solvable. Burnside, in the first edition of his book [18], had studied simple groups of odd order. In a series of three papers, Miller and Burnside [19, 20, 81] proved that there were no simple groups of odd composite order less than 40,000. In Note M on page 503 of the second edition of his book, Burnside [21] conjectured that there were no simple groups of odd composite order. Equivalent to his conjecture is that all groups of odd order are solvable.

Given a normal subgroup N of a group G with $(|N|, |G/N|) = 1$, the fact that any two complements of N must be conjugate was finally resolved in 1962 by Feit and Thompson [36] after they had proven that any group of odd order is solvable [37]. In the interim, D.G. Higman [59] proved (in 1954) a result concerning when complements are conjugate.

Theorem 2.18. [Theorem 5 in D.G. Higman [59]] *For an extension G over a normal subgroup N of G, such that $|N|$ is relatively prime to its index, the following are equivalent statements.*

(i) *If C and D are complements of N in G, then they are conjugate in $\langle C, D\rangle$.*
(ii) *For each subgroup H of G, such that $G = NH$, and for each pair C and D of complements of $N \cap H$ in H, then there exists an automorphism α of H, such that $C = \alpha(D)$.*

(iii) *For each subgroup H of G, such that $G = NH$, for each complement C of $N \cap H$ in H, and for each prime p, there exists a Sylow p-subgroup S of $N \cap H$, such that C is part of the normalizer of S.*

D.G. Higman's Proof. (i)⇒(ii) Lemma 1.11 implies $G = [N]C = [N]D$. By (i), there exists an element $x \in \langle C, D\rangle \leq H$, such that $D^x = C$. The result follows as conjugation by x induces an automorphism of H.

(ii)⇒(iii) Assume $G = NH$ for a subgroup H of G, with $N \cap H \neq \{1\}$, and let P be a Sylow p-subgroup of $N \cap H$. Since $N \cap H \trianglelefteq H$, The Frattini argument (Lemma 1.7) implies $H = (N \cap H)\mathcal{N}_H(P)$. Since $N \cap H \cap \mathcal{N}_H(P) \trianglelefteq \mathcal{N}_H(P)$ with $(|N \cap H|, [H : N \cap H]) = 1$, Theorem 2.12 implies that $\mathcal{N}_H(P) = [N \cap H \cap \mathcal{N}_H(P)]D$. Then, by Lemma 1.11 it follows that $H = [N \cap H]D$ with $D \leq \mathcal{N}_H(P)$.

Let C be another complement to $N \cap H$ in H. By (ii), there exists an automorphism α of H, such that $C = \alpha(D)$. This implies that C normalizes the Sylow p-subgroup $S = \alpha(P)$ of $N \cap H$.

(iii)⇒(i) Let C and D be two complements of N in G. Proceed by induction on the order of $H = \langle C, D\rangle$. It follows from Lemma 1.11 that $H = [N \cap H]C = [N \cap H]D$. First, assume that $N \cap H$ is nilpotent. Given that $(|N \cap H|, [H : N \cap H]) = 1$, it follows from Theorem 2.13 that there exists an element $h \in H$, such that $C = D^h$ and C and D are conjugate in H.

Now assume that $N \cap H$ is not nilpotent. Thus, there is a Sylow p-subgroup P of $N \cap H$, such that $\mathcal{N}_H(P) \neq H$. By (iii), there exists Sylow p-subgroups Q and R of $N \cap H$, such that Q is normalized by C and R is normalized by D. There exists an element $x \in H$, such that $Q = R^x$. By Corollary 1.13, the subgroup D^x is a complement to N in G. Given that $Q^{x^{-1}dx} = R^{dx} = R^x = Q$, for all $d \in D$, D^x normalizes Q. This implies that $\langle C, D^x\rangle \leq \mathcal{N}_H(P) < H$. By induction, there exists an element $y \in \langle C, D^x\rangle$, such that $C = D^{xy}$. Given that both x and y are elements of $\langle C, D\rangle$, the element $xy \in H$ and result (i) follows. □

D.G. Higman used Theorem 2.18 to show that Zassenhaus's conjecture was equivalent to the following statement.

Statement 2.19. [(+) in D.G. Higman [59]] *If G is a group and Δ a group of automorphisms of G, such that the orders of Δ and G are relatively prime, then for each prime p, there exists a Sylow p-subgroup of G which is mapped onto itself by every automorphism in Δ.*

Now comes one of the biggest theorems in finite group theory. In the early 1960s, Feit and Thompson proved the famous odd order theorem.

Theorem 2.20. [The odd order theorem in Feit and Thompson [37]] *Any finite group of odd order is solvable.*

This result established the form of the Schur-Zassenhaus theorem as it is known by today.

Theorem 2.21. [The Schur-Zassenhaus Theorem or Theorem 2 in Feit and Thompson [36]] *If N is a normal subgroup of a group G and if the order of N and the order of the index of N in G are relatively prime, then there is a complement to N in G and any two such complements are conjugate.*

Note that in Theorem 2.21, N is a Hall π-subgroup of the group G. In addition, recall that all normal subgroups are also permutable. Using these two facts, a number generalizations of the Theorem 2.21 (the Schur-Zassenhaus theorem) have been obtained by requiring that a Hall π-subgroup satisfy certain permutability conditions.

The first such result was published by Šemetkov in 1968. Before he presented his result, he indicates that he is generalizing "Schur's theorem," by which he means Theorem 2.12.

Theorem 2.22. [Theorem 3 in Šemetkov [98]] *Let G be a group, and let π be a set of primes. If the order of $O^{\pi}(G)$ is not divisible by p^2 for any $p \in \pi$, then $O^{\pi}(G)$ has at least one complement in G.*

If $O^{\pi}(G)$ is itself a π'-subgroup, the result is essentially Theorem 2.12.

Beweis. Given that $K = O^{\pi}(G)$ is not contained in $\Phi(G)$, there exists a proper subgroup H of G, such that $G = KH$ is a reduced product. By Lemma 1.60, $K \cap H \leq \Phi(H)$.

Suppose that $p \big| |K \cap H|$, where $p \in \pi$. Let P_1 be a Sylow p-subgroup of $K \cap H$. Note that by the hypothesis $|P_1| = p$ and P_1 is also a Sylow p-subgroup of K. Since $K \cap H \lhd H$, it follows from the Frattini Argument (Lemma 1.7) that $H = \mathcal{N}_H(P_1)(K \cap H) = \mathcal{N}_H(P_1)$ as $K \cap H \leq \Phi(H)$. This implies $P_1 \lhd H$.

Now $P_1 \leq P$, where P is the Sylow p-subgroup of H with $|P| = p^{\alpha}$ for $\alpha \geq 1$. Suppose that P is not a Sylow p-subgroup of G. Then, $P < S$, where S is a Sylow p-subgroup of G with $|S| = p^{\beta}$ for $\beta > \alpha$. Since $G = HK$, it follows that $|G| = |H||K|/|H \cap K|$. Given that p^{α} is the highest power of p that divides $|H|$ and p^1 is the highest power of p that divides $|K \cap H|$, there must be a power of p greater than or equal to p^2 that divides $|K|$. This contradiction forces P to be a Sylow p-subgroup of G. Given that $P \leq H$ and $P_1 \lhd H$, it follows that $P_1 \lhd P$. The fact that $|P_1| = p$ implies that P_1 is elementary abelian and that $P_1 \leq Z(P)$.

Now suppose that $p \big| |K \cap H|$, where $p \notin \pi$. Let Q_1 be the Sylow p-subgroup of $K \cap H$. By the same argument as presented above, it follows that $Q_1 \lhd H$. Suppose that $Q_1 < Q$, where Q is a Sylow p-subgroup of H. Then, $|Q_1| = p^{\alpha}$ and $|Q| = p^{\beta}$ with $\alpha < \beta$. However, $|G|/|K| = |H|/|K \cap H|$ or $[G : K] = |H|/|K \cap H|$ implies that $p \big| [G : K]$, a contradiction. Thus, Q_1 is a Sylow p-subgroup of H with $Q_1 \lhd H$. However, $Q_1 \leq \Phi(H)$, which is also a contradiction. Thus, for each prime p, such that $p \big| |K \cap H|$, it must be that $p \in \pi$.

In summation, $G = KH$ and for each prime p that divides $|H \cap K|$, there is a Sylow p-subgroup P of G, such that $P \cap K$ is elementary abelian and contained in the center of P. This implies by Theorem 2 in [98] that there exists a subgroup $N \lhd G$, with $N \leq K$, such that $G = [N]H$. Proceed by induction on the order of G. First, note that $K \cap H = O^{\pi}(G) \cap H = O^{\pi}(H)$. Since $O^{\pi}(H) = K \cap H$ and is not divisible by p^2 for any $p \in \pi$, there is a subgroup L of H, such that $H = [K \cap H]L$. Given that $K \cap H \leq \Phi(G)$, this is a contradiction unless $K \cap H = \{1\}$ and $H = L$. This implies $G = [K]H$ and that K has a complement in G. □

Šemetkov also notes that Theorem 2.22 "remains valid if the subgroup K satisfies the following weaker condition: K contains a Sylow p-subgroup, $p \in \pi$, which is abelian for $p = 2$, modular for $p > 2$ and the order of K is not divisible by q^2 for any $q \in \pi$, $q \neq p$" (page 166 of [98]).

Before the next generalization is given, the following definition and lemma are required. The proof of the lemma can be found in [27].

Definition 2.23. A subgroup A of a group G is **s-seminormal** if A permutes with all Sylow p-subgroups of G with $(p, |A|) = 1$.

Lemma 2.24. [Lemma 1 in Chen [27]] *Let H and K be subgroups of a group G and $\pi = \pi(H)$. Then*
(i) *Assume H is s-seminormal in G. If $H \leq K$ or $H \cap K$ is a Hall π-subgroup of G, then $H \cap K$ is s-seminormal in K.*
(ii) *If H is seminormal in G, then $H \cap K$ is s-seminormal in K.*

Theorem 2.25. [Theorem 1 in Chen [27]] *If a Hall π-subgroup H of a group G is s-seminormal in G, then H has a complement in G and all such complements are conjugate in G.*

Chen's Proof. Proceed by induction on the order of G. Let P be a Sylow p-subgroup of G, such that $p \notin \pi$. If $G = HP$, the result follows by Sylow. Assume that $HP < G$. Since $HP^g = P^gH$ for all $g \in G$, then H or P is contained in a proper normal subgroup N of G (VI.4.10 in [66]).

Case 1: $H \leq N$. If $H = N$, the results follows by Theorem 2.21. Assume that $H < N$. By Lemma 2.24, H is s-seminormal in N. Thus, there exists a subgroup K in N, such that $N = HK$ with $H \cap K = \{1\}$ and all complements to H in N are conjugate in N.

Now consider $\mathcal{N}_G(K)$. If $\mathcal{N}_G(K) = G$, then K is normal in G. Given that $N/K \cong H$ and $N/K \lhd G/K$, N/K is a Hall π-subgroup of and s-seminormal in G/K. By induction, N/K has a complement K_1/K in G/K. It follows that $G = NK_1 = HKK_1 = HK_1$ with $H \cap K_1 = \{1\}$. If K_2 is another complement to H in G, then, as $K \leq K_2$, K_1/K and K_2/K are conjugate in G/K. This implies by Lemma 1.14 that K_1 and K_2 are conjugate in G.

If $\mathcal{N}_G(K) < G$, consider the subgroup $H\mathcal{N}_G(K)$ of G, and let $g \in G$. Then, $K^g \le N$ and K^g is a complement of H in N. Then, there exists an element $n = kh \in N$, where $k \in K$ and $h \in H$, such that $K^g = K^{kh} = K^h$. This implies $K^{gh^{-1}} = K$ and that $gh^{-1} \in \mathcal{N}_G(K)$. It follows that $G = H\mathcal{N}_G(K)$.

Let $D = H \cap \mathcal{N}_G(K)$. Given that

$$|G| = |H\mathcal{N}_G(K)| = |H||\mathcal{N}_G(K)|/|H \cap \mathcal{N}_G(K)|$$

or

$$[G : H] = [\mathcal{N}_G(K) : D],$$

D is a Hall π-subgroup of $\mathcal{N}_G(K)$. By Lemma 2.24, D is s-seminormal in $\mathcal{N}_G(K)$ and $\mathcal{N}_G(K) = DK_1$, where $D \cap K_1 = \{1\}$ and K_1 is a π'-group. Furthermore, all complements of D in $\mathcal{N}_G(K)$ are conjugate. In addition, $G = H\mathcal{N}_G(K) = HDK_1 = HK_1$, with $H \cap K_1 = \{1\}$, and K_1 is a complement of H in G. Let K_2 be another complement of H in G. Then, K_2 is a Hall π'-subgroup of G with $N \cap K_2 = K_3 \lhd K_2$. Furthermore,

$$HK_3 = H(K_2 \cap N) = HK_2 \cap H = N,$$

with $H \cap K_3 = H \cap K_2 \cap N = \{1\}$. Thus, K_3 is a complement to H in N, which means K_3 and K are conjugate in N. This implies $\mathcal{N}_G(K_3)$ and $\mathcal{N}_G(K)$ are conjugate. Since $K_2 \le \mathcal{N}_G(K_3)$, it follows that $K_2^g \le \mathcal{N}_G(K)$ for some $g \in G$. This means $\mathcal{N}_G(K) = DK_2^g$ and that K_2^g and K_1 are conjugate in $\mathcal{N}_G(K)$. Thus, K_1 and K_2 are conjugate in G.

Case 2: $H \nleq N$. The subgroup HN/N is a Hall π-subgroup of G/N by Lemma 1.27. Let P/N be a Sylow p-subgroup of G/N with $p \in \pi'$. Then, $P = P_1N$ for some Sylow p-subgroup P_1 of G. Since H is s-seminormal in G, $HP_1 \le G$ and $HP_1N = HP \le G$. This implies $HP/N = H/N \cdot P/N$ is a subgroup of G/N. Thus, HN/N is s-seminormal in G/N. By induction, HN/N has a complement K/N in G/N and all complements of HN/N in G/N are conjugate. It follows that $G = HK$ with $H \cap K = D$, where D is a Hall π-subgroup of K. Lemma 2.24 implies D is s-seminormal in K. By induction, $K = DK_1$, where $D \cap K_1 = \{1\}$, and all complements of D in K are conjugate. It follows that $G = HK_1$, where $H \cap K_1 = \{1\}$. Let K_2 be another complement to H in G. Since K_1N/N is a complement to HN/N in G/N, $K_1^xN = K$ for some $x \in g$. The same argument results is a $y \in G$, such that $K_2^yN = K$. Since $G = HK_1^x = HK_2^y$, the modular identity (Lemma 1.6) implies

$$K = G \cap K = K_1^x H \cap K = K_1^x(H \cap K) = DK_1^x$$

and

$$K = G \cap K = K_2^y H \cap K = K_2^y(H \cap K) = DK_2^y.$$

It then follows by induction that there is a $z \in K$, such that $K_1^{xz} = K_2^y$ or that $K_1^{xzy^{-1}} = K_2$. □

The final series of results presented in this section are given without proof due to the amount of machinery needed for formal proofs. The following definition, generalizing the concept of a permutable subgroup, is needed for the next generalizations of the Schur-Zassenhaus theorem.

Definition 2.26. Let A and B be subgroups of a group G and let X be a nonempty subset of G. The subgroup A is **X-permutable** with B if $AB^x = B^xA$ for some $x \in X$.

More information on X-permutability can be found in [48]. It should be noted that in Theorems 2.27, 2.28, 2.29, and 2.30 that the Hall subgroup being complemented need not be normal in the group.

Theorem 2.27. [Theorem 1.2 in Guo et al. [49]] *Let A be a Hall subgroup of a group G and $X = \mathcal{F}(G)$ the Fitting subgroup of G. Suppose that A is X-permutable with all subgroups of some supplement T of A in G. Assume also that either A or T is soluble. Then, A is complemented in G and any two of its complements are conjugate in G.*

Theorem 2.27 was immediately generalized three years later.

Theorem 2.28. [Theorem 3.1 in Guo and Skiba [50]] *Let X be a normal π-separable subgroup of G and A a Hall π-subgroup of G. Let $G = AT$ for some subgroup T of G, and let q be a prime. If A is X-permutable with every Sylow p-subgroup of T for all primes $p \neq q$ and either A is soluble or every π'-subgroup of T is soluble, then T contains a complement of A in G and any two complements of A in G are conjugate.*

Letting $X = \{1\}$, the following result is obtained.

Theorem 2.29. [Theorem A* in Guo and Skiba [50]] *Let A be a Hall π-subgroup of a group G. Let $G = AT$ for some subgroup T of G, and let q be a prime. If A permutes with every Sylow p-subgroup of T for all primes $p \neq q$, then T contains a complement of A in G and any two complements of A in G are conjugate.*

Theorem 2.30. [Theorem 3.7 in Li and Zhang [77]] *Let X be a soluble normal subgroup of a group G and H a Hall π-subgroup of G. If H is X-permutable with all Sylow subgroups of some supplement T of H, then H has a complement and all its complements are conjugate in G.*

3 Abelian and minimal normal subgroups

In 1902, Schur showed (Theorem 2.7) that if N is a subgroup of a group G such that $N \leq Z(G)$ and $(|N|, |G/N|) = 1$, then there exists a subgroup H of G such that $G = N \times H$. Under this very specialized condition on the subgroup N of group G, Schur was able to show N had a complement in G. One avenue in the normal subgroup complementation investigation was to weaken one or both of the conditions established by Schur for when N has a complement in G. Some of the veins of this investigation were presented in Chapter 2.

Schur's condition that the subgroup N of a group G satisfy $N \leq Z(G)$ means that N is abelian. One natural next step would be to establish conditions for when an abelian normal subgroup has a complement. This avenue was traveled by numerous group theorists and the results they established are presented in this chapter. Conditions for when a minimal normal subgroup has a complement are also included in this chapter for the obvious reasons. As demonstrated by Lemma 1.61, which is restated here, very simple conditions can be established for when an abelian minimal normal subgroup N of a group G has a complement in G.

Lemma 3.1. *Let N be an abelian minimal normal subgroup of a group G. If $N \nleq \Phi(G)$, then $G = [N]M$ for some maximal subgroup M in G.*

The following result follows directly from Lemma 3.1

Corollary 3.2. *Let N be a minimal normal subgroup of a solvable group G. If $N \nleq \Phi(G)$, then $G = [N]M$ for some maximal subgroup M in G.*

Right after Zassenhaus published his complementation theorems (Theorems 2.12 and 2.13) in 1937 advancing Schur's 1902 result, Ore [84–86] wrote a series of seminal papers where "the guiding principle is the application of structure theory to the theory of groups" (page 431 in [86]). In these papers, he studied groups through quotients, permutable subgroups, maximal subgroups, group decompositions, and subgroup series.

To start, Ore states and proves in Chapter IV of [86] on solvable groups a slightly different version of Lemma 1.61.

Theorem 3.3. [Theorem IV.4 in Ore [86]] *Let M be a maximal subgroup of a finite group G and suppose G contains a minimal abelian normal subgroup L of order p^α. If L is not contained in M, then $G = [L]M$ and M is normal or has p^α conjugates.*

Note that if M is not normal in G, then $\mathcal{N}_G(M) = M$, which implies that M has p^α conjugates. Each one of these conjugates, by Corollary 1.13, is a complement to L in G.

Ore included the remark on M being normal or having p^α conjugates due to his interest in maximal subgroups. One of his goals in Chapter IV of [86] was to

DOI 10.1515/9783110480214-003

establish the existence of certain types of maximal subgroups in solvable groups with the importance of maximal subgroups having been already established by his Theorem IV.4 (Theorem 3.3 above). The next two results listed here are by Ore and are part of his investigation in [86] of maximal subgroups. They are of interest here as they establish splitting criteria for certain abelian normal subgroups. Before proceeding, the following definition (from [86]) is needed.

Definition 3.4. Let H be a subgroup of some group G. The subgroup H is **decomposably contained** in G if there exist two normal subgroups N and N^* of G with $N \leq N^*$ such that $G = N^*H$ and $N^* \cap H = N$. The group G **splits regularly** over N^*/N if any other decomposition $G = N^*H_1$ with $N^* \cap H_1 = N$ implies that H and H_1 are conjugate in G.

Essentially, if a subgroup H of a group G is decomposably contained in G, then $G/N = [N^*/N]H/N$ for normal subgroups N and N^* of G such that $N \leq N^*$. The motivation for this concept comes from his investigation of solvable groups.

Theorem 3.5. [Theorem IV.6 in Ore [86]] *Let $B \trianglelefteq A$ and let A split regularly over B, where $A = [B]H$, for $H \leq A$, such that $\mathcal{N}_A(H) = H$. Then, any group G in which A and B are contained normally will also split regularly where $G = [B]N$, $N = \mathcal{N}_G(H)$, and $\mathcal{N}_G(N) = N$.*

Ore's Proof. Since $H < A \trianglelefteq G$, for each $g \in G$, there is an $a \in A$ such that $H^g = H^a$. Thus, $H^{ga^{-1}} = H$ and $ga^{-1} \in N = \mathcal{N}_G(H)$. This implies $ga^{-1} = n$ or $g = na$ for some $n \in N$. Since $A \trianglelefteq G$, it follows that $g = a'n$ for $a' \in A$. Given that $A = [B]H$, $g = a'n = (bh)n = b(hn)$ for $b \in B$, $h \in H$, and $hn \in \mathcal{N}_G(H)$. This means $G = BN$. Given that $\mathcal{N}_A(H) = H$ and $A = [B]H$, it follows that $B \cap N = \{1\}$ and that $G = [B]N$. The fact that $G = [B]N$ means that $\mathcal{N}_G(N) = N$.

Suppose $G = [B]M$ for another subgroup M of G. By Lemma 1.11, $A = [B](A \cap M)$, where $A \cap M$ and H are conjugate in A. Thus, there is an element $a \in A$ such that $H^a = A \cap M$. Clearly, $A \cap M \lhd M$. If $M < \mathcal{N}_G(A \cap M)$, there exists an element $b \in B$ such that $(A \cap M)^b = A \cap M$ or $H^{ab} = H^a$. This means $aba^{-1} \in N$ or that $b \in a^{-1}Na$. This is a contradiction by Corollary 1.13. As a result, $\mathcal{N}_G(A \cap M) = M$.

Now consider the subgroup N^a. For each $x \in N^a$, $x = a^{-1}na$ for $n \in N$, and

$$(A \cap M)^x = (H^a)^{a^{-1}na} = H^{na} = H^a = A \cap M$$

and $x \in \mathcal{N}_G(A \cap M)$. This means $M = N^a$ and that M and N are conjugate in G. □

Theorem 3.5 appears to be the first reduction theorem for splitting over a normal subgroup. It ties in with Ore's study of maximal subgroups as any non-normal maximal subgroup M of a group G satisfies $\mathcal{N}_G(M) = M$. Furthermore, Theorem 3.5 can be used to establish the existence, under certain conditions, of a maximal subgroup. Let G be a group with a unique minimal normal subgroup B such that B is an elementary abelian p-group. Now consider a subgroup A of G such that $\{1\} \lhd B \lhd A \lhd \cdots$ is part of a normal series for G with A/B an elementary abelian q-group with $q \neq p$. It is easy to show that $A = [B]C$ such that C is a Sylow q-subgroup of A with $\mathcal{N}_A(C) = C$. The conditions of Theorem 3.5 are now satisfied

Ore uses Theorem 3.5 to prove the following result which also gives a condition for the existence of a certain type of maximal subgroup. The result, listed below, is included in this section as it is also an abelian normal subgroup complementation theorem.

Theorem 3.6. [Theorem IV.7 in Ore [86]] *Let group G contain normal subgroups A and B such that $B \leq A$, where B is a unique minimal normal abelian subgroup of G of order p^α and elementary abelian, while A/B is some minimal normal abelian subgroup of G/B of order q^β and elementary abelian with $p \neq q$. Then, $A = [B]C$, where $C \cong A/B$ and $\mathcal{N}_A(C) = C$. The group G also splits over B with $G = [B]M$, where $M = \mathcal{N}_G(C)$. Furthermore, M is maximal in G and the number of its conjugates is $k = p^\alpha \equiv 1 (\mathrm{mod}\ q)$.*

Ore gets his maximal subgroup and a new splitting theorem is also obtained. In a paper published in 1952, Gaschütz restates Ore's Theorem 3.6 in a slightly more general form. Gaschütz's more general result follows from a theorem he proves earlier in his paper [41], but it is essentially Ore's result and he does give credit to Ore.

Theorem 3.7. [Satz 6 in Gaschütz [41]] *If B is a unique minimal normal subgroup of a group G and if G/B contains a nontrivial abelian normal subgroup A/B such that $([A : B], |B|) = 1$, then G splits over B.*

The next stop on this journey of abelian normal subgroup complementation is Gaschütz's seminal 1952 paper on normal subgroup complementation. Using extension theory, he proved many new splitting theorems and generalized a few known ones. More specifically, Gaschütz's goal was to investigate extensions over abelian normal subgroups and to extend results that had been proven by Bergström, Ore, Schur, and Zassenhaus.

His first non-reduction result concerning splitting over an abelian normal subgroup generalizes two results by Bergström (Satz 1 and Satz 2 in [16]). The proof is omitted.

Theorem 3.8. [Satz 3 in Gaschütz [41]] *Let A be an abelian normal subgroup of a group G. If G/A contains a normal nilpotent subgroup T such that no elements of A, other than the identity element, remain invariant under T, then G splits over A.*

Many more results from Gaschütz's 1952 paper will appear in this book. Just a year later in 1953, Gaschütz [42] published a paper where he proves a variety of fundamental results concerning the Frattini subgroup. This paper also contains a generalization of Lemma 1.61, which is of interest here.

Theorem 3.9. [Satz 7 in Gaschütz [42]] *If A is an abelian normal subgroup of a group G and $A \cap \Phi(G) = \{1\}$, then G splits over A. Moreover, A is the direct product of minimal normal subgroups of G.*

Beweis. If A is a minimal normal subgroup of G, then G splits over A by Lemma 1.61. Suppose that A is not minimal normal in G and proceed by induction on the order of G. Let $N < A$ be a minimal normal subgroup of G. Then, by Lemma 1.61, $G = [N]M$ for M a maximal subgroup of G. Since A is abelian, $B = A \cap M$ is normal in G and is an abelian normal subgroup of M.

Given that $G = [N]M$ with A abelian, $B \cap \Phi(M)$ is normal in G. By Theorem 1.53, $B \cap \Phi(M) \leq \Phi(G)$. Since $A \cap \Phi(G) = \{1\}$, this implies $B \cap \Phi(M) = \{1\}$. Thus, $M = [B]H$, for $H < M$, and B is the direct product of minimal normal subgroups of M. Given that $G = NM$, with B abelian, each of these minimal normal subgroups in M are minimal normal subgroups in G. By Lemma 1.11, $G = [A]H$, with $A = N \times B$. □

Corollary 3.10. *If $\Phi(G) = \{1\}$ for a group G, then G splits over every abelian normal subgroup.*

Gaschütz used Theorem 3.9 to show that if A is an abelian normal subgroup of a group G, then $R_G(A) \leq \Phi(G)$, where $R_G(A)$ is the intersection of those subgroups normal in G and maximal in A. As a interesting note, $R_G(G) = \Phi(G)$ if and only if G is nilpotent.

Definition 3.11. Let G be a group. The **sockel** of a group G or $\mathrm{Soc}(G)$ is the product of the minimal normal subgroups of G.

One of the properties of $\mathrm{Soc}(G)$ is that $\mathrm{Soc}(G) = S_A \times S_N$, where S_A, the **abelian component of the sockel**, is generated by the abelian minimal normal subgroups of G and S_N, the **non-abelian component of the sockel**, is generated by the non-abelian minimal normal subgroups of G (see [96] for more information on the sockel of a group). Gaschütz [42] proves the following result.

Theorem 3.12. [Satz 14 in Gaschütz [42]] *A group G satisfies $\Phi(G) = \{1\}$ if and only if G splits over the abelian component of its sockel.*

Beweis. The condition is sufficient by Theorem 3.9. To show the condition is necessary, suppose that $G = [S_A]H$, where S_A is the abelian component of the sockel and $H \leq G$. If $\Phi(G) \neq \{1\}$, there is a minimal normal subgroup N of G such that $N \leq \Phi(G)$. Since N is abelian, $N \leq S_A$. There exists a normal subgroup M of G such that $S_A = N \times M$.

By Lemma 1.11, $G = [N]MH$, which is a contradiction as $N \leq \Phi(G)$. This implies $\Phi(G) = \{1\}$. □

Note that when G is a solvable group with $\Phi(G) = \{1\}$, the Fitting subgroup $\mathcal{F}(G)$ is the abelian component of the sockel of G. Thus, Theorem 3.12 motivates the following theorem by Bechtell.

Theorem 3.13. [Theorem 6.1.4 in Bechtell [7]] *Let G be a finite solvable group with $\Phi(G) = \{1\}$ and $\mathcal{F}(G) \neq \{1\}$. Then, $G = [\mathcal{F}(G)]C$ for $C \leq G$ such that*
(i) *$C \cong C^* \leq Aut(\mathcal{F}(G))$ and*
(ii) *if a Sylow p-subgroup P of $\mathcal{F}(G)$ has order p^n, then C is homomorphic to a subgroup of the general linear group $GL(k,p)$ for some $k \leq n$.*

Bechtell's Proof. By Corollary 1.54, $\Phi(\mathcal{F}(G)) \leq \Phi(G) = \{1\}$ and $\mathcal{F}(G)$ is elementary abelian. Thus, by Theorem 3.9, $G = [\mathcal{F}(G)]C$ for $C \leq G$.

To prove (i), first note by Theorem 1.69 that $C_G(\mathcal{F}(G)) \leq \mathcal{F}(G)$. This implies, given that $\mathcal{F}(G)$ is abelian, that $C_G(\mathcal{F}(G)) = \mathcal{F}(G)$. Since each element $c \in C$ induces (via conjugation) an isomorphism of $\mathcal{F}(G)$, the homomorphism $\phi : C \to Aut(\mathcal{F}(G))$ has a trivial kernel. This means $C \cong C^* \leq Aut(\mathcal{F}(G))$.

For (ii), let P be a Sylow p-subgroup of $\mathcal{F}(G)$ of order p^n with $n \geq 1$. Since $P \trianglelefteq G$, there exists a minimal normal subgroup N of G such that $N \leq P$ with $|N| = p^k$, where $k \leq n$. Given that $Aut(N) \cong GL(k,p)$ and $N \trianglelefteq G$, there is a homomorphism $\psi : C \to Aut(N)$. Thus, C is isomorphic to a subgroup of $GL(k,p)$. □

Baer [1] published a paper in 1953 that investigated nilpotent characteristic subgroups of finite groups. In this paper, he proved many fundamental results concerning the Frattini subgroup and the Fitting subgroup. There are a number of results from that paper that are of interest here as they lead to a splitting result by Gaschütz. The first is a result that is essentially Lemma 1.61.

Lemma 3.14. [Lemma 2.1 in Baer [1]] *The following properties of an abelian minimal normal subgroup N of a group G and the subgroup M are equivalent.*
(i) *M is a complement of N in G.*
(ii) *M is a maximal subgroup of G with $N \nleq M$.*
(iii) *$G = NM$ and $M \neq G$.*
(iv) *$G = NM$ and $N \nleq M$.*

After proving this result, Baer states the well-known fact that the subgroup N must be elementary abelian. Lemma 3.14 is not true if N is not abelian. For example, take your favorite symmetric group $\mathcal{S}_n$ for $n \geq 5$.

A section of Baer's 1953 paper is dedicated to the automorphisms of minimal normal subgroups. He first proves that for a group G, the Fitting subgroup $\mathcal{F}(G)$ centralizes every minimal normal subgroup of G. He then makes the following observation.

Let N be a normal subgroup of a group G. Then, $G/C_G(N)$ is isomorphic to the group of automorphisms of N induced by the elements of G. If S is a subgroup of G such that $G = C_G(N)S$, then every automorphism of N induced by an element of G is induced by an element in S. This motivates the following definition by Baer.

Definition 3.15. The group G is **represented in its normal subgroup N by its subgroup S** whenever $G = C_G(N)S$.

Since a subgroup S will always exist with the above property (take $S = G$), a minimal subgroup S representing G will always exist. Baer then proceeded to prove the following three results concerning minimal normal subgroups. The lemmas are straightforward and presented without proof. While not providing any new splitting criteria, they do lead to a nice complementation result by Gaschütz.

Lemma 3.16. [Lemma 4.1 in Baer [1]] *Assume that N is a minimal normal subgroup of a group G and G is represented in N by its subgroup S, then*
(i) $N \leq S$ *or* $N \cap S = \{1\}$,
(ii) $S \cap C_G(N)$ *is a normal subgroup of* NS, *and*
(iii) N *is minimal normal in* NS.

Lemma 3.17. [Lemma 4.2 in Baer [1]] *Assume that the minimal normal subgroup N of G is abelian and that S is a minimal subgroup representing G in N. Then*
(i) NS *is a minimal subgroup T with the properties* $N \leq T$ *and* $G = TC_G(N)$,
(ii) $S \cap C_G(N) \leq \Phi(S)$, *and*
(iii) $\mathcal{F}(NS) = N(C_G(N) \cap S) = C_G(N) \cap NS$.

Theorem 3.18. [Proposition 4.2 in Baer [1]] *If N is a minimal normal subgroup of a group G such that $G/C_G(N)$ contains a nontrivial normal subgroup whose order is relatively prime to $|N|$, then*
(i) N *is abelian,*
(ii) *every minimal S representing G in N satisfies* $S \cap N = \{1\}$ *and* $S \cap C_G(N) = \Phi(SN)$, *and*
(iii) *the two minimal subgroups H and K representing G in N satisfy* $NH = NK$ *if and only if there exists* $x \in N$ *such that* $H = x^{-1}Kx$.

Baer's Proof. Let $Q/C_G(N)$ be the normal subgroup of $G/C_G(N)$ such that $(|Q/C_G(N)|, |N|) = 1$. To prove (i), suppose that $M \cap C_G(N) = \{1\}$. This implies $|N| = [NC_G(N) : C_G(N)]$ and that

$$([NC_G(N) : C_G(N)], [Q : C_G(N)]) = 1.$$

Thus, as subgroups of $G/C_G(N)$, $NC_G(N)/C_G(N) \cap Q/C_G(N) = \{1_{G/C_G(N)}\}$, and by the modular identity (Lemma 1.6), $C_G(N) = NC_G(N) \cap Q = C_G(N)(N \cap Q)$. This means

$Q \cap N \leq C_G(N)$. The assumption of $N \cap C_G(N) = \{1\}$ forces $Q \cap N = \{1\}$. However, this implies, since both Q and N are normal in G, that $Q \leq C_G(N)$ is a contradiction. Thus, $N \cap C_G(N) \neq \{1\}$. Given that N is minimal normal in G, $N \cap C_G(N) = N$ or $N \leq C_G(N)$. Thus, N is abelian.

Before proving the first of the two statements in (ii), note that N is an elementary abelian p-group for prime p. Now let S be minimal such that $G = C_G(N)S$ and let $J = S \cap C_G(N)$. By Lemmas 3.16 and 1.60, $J \lhd NS$ with $J \leq \Phi(S)$. Furthermore, Theorem 1.53 implies that $J \leq \Phi(NS)$.

Let $N_1 = Q \cap S$, where $J \leq N_1$. Then, by the modular identity (Lemma 1.6)

$$N_1 C_G(N) = C_G(N)(Q \cap S) = C_G(N)S \cap Q = G \cap Q = Q$$

and

$$N_1/J = (S \cap Q)/(S \cap C_G(N)) \cong (S \cap Q)C_G(N)/C_G(N) = Q/C_G(N),$$

which implies that $[N_1 : J] \neq 1$ and $([N_1 : J], p) = 1$.

Let r be the greatest divisor of $|N_1|$ such that $(r, p) = 1$. In other words, $|N_1| = rp^\alpha$ with $(r, p) = 1$ and $\alpha \geq 1$. Then, $r = [N_1 : J]l$ for some positive integer l. Since $J \leq \Phi(S)$, J is nilpotent. Therefore, $J = \overline{P} \times P$, where $|\overline{P}| | r$ and $|P|$ is a power of p. Furthermore, $P \trianglelefteq N_1$ with $[N_1 : P] = [N_1 : J][J : P]$. Since $[N_1 : J] | r$, it follows that $[J : P] = |\overline{P}|$, $(|P|, [N_1 : P]) = 1$, with $|P| = p^\alpha$. By Theorem 2.12, $N_1 = [P]T$, with $|T| = r$. By Lemma 1.10, $T \lhd S$ and $N_1 = P \times T$.

Assume that $N \cap S \neq \{1\}$. Then, by Lemma 3.16, $N \leq S$. Since N is abelian, $N \leq C_G(N)$ and $N \leq C_G(N) \cap S \leq J$. Given that P is the collection of all elements of N_1 of order a power of p, $N \leq P \leq J \leq C_G(N)$. However, since $N_1 = P \times T$, it follows that $N_1 = P \times T \leq C_G(N)$. This means

$$N_1 = N_1 \cap C_G(N) = S \cap Q \cap C_G(N) = S \cap C_G(N) = J,$$

which is a contradiction. This implies $N \cap S = \{1\}$.

The second statement in (ii) will be proven after (iii) is proven. However, continuing in the same vein, consider the subgroup $NS \cap Q = N(S \cap Q) = NN_1$, which is normal in NS. Since $N_1 = P \times T$ and $P \leq J \leq C_G(N)$, P commutes element-wise with both N and T. This implies N and P are normal subgroups of NN_1 and that $NP \trianglelefteq NN_1$. By Lemma 1.11, it follows that $NN_1 = [NP]T$ and that NP is the collection of all elements of NN_1 of order a power of p.

Recall that $T \lhd S$. Thus, $S \leq \mathcal{N}_{NS}(T)$. Let $a \in \mathcal{N}_{NS}(T)$. Then, $a = ns$ for $n \in N$ and $s \in S$. Since both a and s are in $\mathcal{N}_{NS}(T)$, it follows that $n \in \mathcal{N}_{NS}(T)$. If $x \in T$, then $[x, n] \in T \cap N = \{1\}$. Thus, n commutes with every element of T. Since $N_1 = P \times T$, $P \leq C_G(N)$, and $Q = N_1 C_G(N)$, the element n commutes with every element of Q.

Let $N^* = \{x \in N \mid qx = xq \text{ for all } q \in Q\}$. Clearly, N^* is a nonempty ($n \in N^*$) subgroup of G. Furthermore, $N^* < N$, as otherwise would imply $Q \leq C_G(N)$. However, since $Q \trianglelefteq G$, $N^* \trianglelefteq G$, which is a contradiction. This implies $n = 1$ and that $a = ns = 1 \cdot s = s \in S$. Consequently, $S \leq \mathcal{N}_{NS}(T)$.

For (iii), first note that the condition is necessary by Lemma 1.12. Assume that H and K are two minimal subgroups representing G in N, such that $NH = NK = W$. As was shown earlier in the proof, the collection of elements of order a power of p forms a characteristic subgroup W_p of W and that there exists complements $\overline{H}$ and $\overline{K}$ to W_p in W such that $H = \mathcal{N}_W(\overline{H})$ and $K = \mathcal{N}_W(\overline{K})$. By Theorem 2.13, there exists an element $w \in W$ such that $\overline{H} = w^{-1}\overline{K}w$. This implies $H = w^{-1}Kw$. Given that $w \in W = KN$, there exist elements $k \in K$ and $x \in N$ such that $w = kx$. This means

$$H = w^{-1}KW = x^{-1}k^{-1}Kkx = x^{-1}Kx.$$

To prove the second part of (ii), consider the subgroup NS. By Lemma 3.16, N is a minimal normal subgroup of NS. Furthermore, it was shown that $N \cap S = \{1\}$. It then follows that S is a maximal subgroup of NS with $\Phi(NS) \leq S$. Therefore, by Lemma 3.17,

$$\Phi(NS) \leq S \cap \mathcal{F}(NS) = S \cap NS \cap C_G(N) = S \cap C_G(N) = J.$$

Given that $J \leq \Phi(NS)$, it follows that $\Phi(NS) \leq \Phi(S)$.

Now let R be a maximal subgroup of NS.

Case 1: $N \leq R$.

By the modular identity (Lemma 1.6), $R = NS \cap R = N(S \cap R)$ and $S \cap R$ is a maximal subgroup of S. This implies

$$J \leq \Phi(S) \leq S \cap R \leq R.$$

Case 2: $N \nleq R$.

Since N is abelian, $N \cap R = \{1\}$ and $NS = NR$. Therefore,

$$RC_G(N) = RNC_G(N) = SNC_G(N) = SC_G(N) = G$$

and R represents G in N. If a subgroup X of R represents G in N, then $G = XC_G(N) = XNC_G(N)$ and $XN \leq RN = SN$. Note, by the modular identity (Lemma 1.6), that $XN = SN \cap XN = N(XN \cap S)$. Since $N \leq C_G(N)$,

$$G = XNC_G(N) = (XN \cap S)NC_G(N) = (XN \cap S)C_G(N)$$

and $XN \cap S$ represents G in N. However, given that S is minimal with respect to this property, $XN \cap S = S$ and $S \leq XN$. As a result, $XN \leq S \leq XN$ and $XN = SN = RN$. This means $X = T$ and T is minimal in representing G in N. By (iii), there exists an element $n \in N$ such that $R = n^{-1}Sn$. Consequently,

$$J = n^{-1}Jn \leq n^{-1}Sn = R.$$

Since J is in every maximal subgroup of NS, it follows that $J \leq \Phi(NS)$. This means $\Phi(NS) = J = S \cap C_G(N)$. □

Baer goes on to prove the following result.

Lemma 3.19. [Lemma 5.2 in Baer [1]] *If N is a minimal normal subgroup of a group G and $G/C_G(N)$ contains a nontrivial solvable normal subgroup, then (N is abelian and) $G/C_G(N)$ contains a nontrivial normal subgroup whose order is prime to the order of N.*

Now let N be a minimal normal subgroup of a solvable group G such that $C_G(N) = N$. This implies that N is an elementary abelian p-group and that G is not nilpotent. If G/N contains no proper nontrivial normal subgroups, then $([G : N], |N|) = 1$ as G is not nilpotent. In this case, by Theorems 2.12 and 2.13, N has a complement in G and all complements are conjugate. Now assume that G/N is not simple. Then, by Lemma 3.19, G/N contains a nontrivial normal subgroup whose order is relatively prime to $|N|$. Then, by Theorem 3.18, N has a complement in G and all complements are conjugate. Gaschütz states this result in [43], and for the reasons just mentioned, gives credit to Baer.

Theorem 3.20. [Satz 3.3 in Gaschütz [43]] *If N is a minimal normal subgroup of a solvable group G and $C_G(N) = N$, then N has a complement in G and all complements are conjugate.*

It should also be noted that if a group G satisfies the conditions in Theorem 3.20, then $\Phi(G) = \{1\}$. Since $C_G(N) = N$ and N is minimal normal, it follows by Theorem 1.69 that the Fitting subgroup $\mathcal{F}(G) \leq N$. Thus, $\mathcal{F}(G) = N$ or $\mathcal{F}(G) = \{1\}$. Since $\Phi(G) \leq \mathcal{F}(G)$, it also follows that $\Phi(G) = N$ or $\{1\}$. Since N is complemented in G, this implies that $\Phi(G) = \{1\}$.

Now suppose that N is the unique minimal normal subgroup of a solvable group G such that $N \nleq \Phi(G)$. By Lemma 1.70, $N = C_G(N) = \mathcal{F}(G)$. Consequently, by the Baer/Gaschütz result (Theorem 3.20), N has complement in G and all complements are conjugate. This is summed up in the following result by Carter, Fischer, and Hawkes. It essentially, as just shown, follows from work done by Baer.

Lemma 3.21. [Lemma 2.2 in Carter, Fischer, and Hawkes [25]] *Suppose a solvable group G has a unique minimal normal subgroup N. If $N \nleq \Phi(G)$ then*
(i) *$N = C_G(N) = \mathcal{F}(G)$ and*
(ii) *all complements of N in G are conjugate. Conversely, if $\mathcal{F}(G)$ is a minimal normal subgroup, it is the only one.*

As noted in Chapter 2, Zassenhaus proved (Theorem 2.13) that if N is an abelian normal subgroup of a group G with $([G : N], |N|) = 1$, then N has a complement in G and all complements of N in G are conjugate. In 1954, D.G. Higman, motivated by the work done by Gaschütz in [41], published a paper [59] whose goal was to generalize the Schur/Zassenhaus result (Theorem 2.12) and to address the conjugacy of complements issue raised by Zassenhaus. More on this paper will appear in Chapter 4. Higman proved the following two results on the conjugacy of complements as corollaries to a theorem on extensions and are companion results to Zassenhaus's Theorem 2.13.

Theorem 3.22. [Corollary 1 in D.G. Higman [59]] *Let A be an abelian normal subgroup and H be a subgroup of a group G such that $A \leq H \leq G$. Let $m = [G : H]$. If m is relatively prime to $|A|$, then any two complements C and D of A in G are conjugate if and only if $C \cap H$ and $D \cap H$ are conjugate in H.*

Theorem 3.23. [Corollary 2 in D.G. Higman [59]] *Let G be a splitting extension of an abelian normal subgroup A. If for each prime p there is a Sylow p-subgroup S of G such that any two complements of $S \cap A$ in S are conjugate in S, then any two complements of A in G are conjugate in G.*

G. Higman, whose work was motivated by D.G. Higman and Gaschütz, worked to further generalize the criteria for when an abelian normal subgroup has a complement and when all complements are conjugate. This next result is a variation of Theorem 3.5 (Ore's result) and its proof uses similar techniques as are used in the proof of Theorem 3.5.

Theorem 3.24. [Theorem 1 in G. Higman [60]] *Let N be an abelian normal p-subgroup of a finite group G. Suppose that G has a p'-subgroup X such that NX is normal in G, but N_0X is not, if N_0 is any normal subgroup of G properly contained in N. Then, N is complemented in G and any two complements of N are conjugate.*

Higman's Proof. Let $g \in G$, and let $Y = g^{-1}Xg$. Given that $NX \trianglelefteq G$, $Y \leq NX$ and $(|Y|, |N|) = 1$. This implies that Y is also a complement of N in NX. Thus, by Theorem 2.13, there exists an element $n \in N$ and $x \in X$ such that $g^{-1}Xg = Y = n^{-1}x^{-1}Xxn = n^{-1}Xn$. This implies $gn^{-1} \in \mathcal{N}_G(X)$, which means $g \in \mathcal{N}_G(X)N$ and $G = N\mathcal{N}_G(X)$.

The group G/N and NX/N act on N. Since NX/N is a p'-group, it follows from Theorem 5.2.3 in [46] that $N = S \times T$, where $S = C_N(X)$ and $T \lhd G$. Clearly, $S \leq N \cap \mathcal{N}_G(X)$. Since $N \cap \mathcal{N}_G(X)$ and X are both normal in $\mathcal{N}_G(X)$ with relatively prime orders, they commute element-wise. As a result, $N \cap \mathcal{N}_G(X) \leq C_N(X) = S$ and $S = N \cap \mathcal{N}_G(X)$. This implies $TX \leq C_G(S)$ and that $TX \lhd ST\mathcal{N}_G(X) = N\mathcal{N}_G(X) = G$. The hypothesis implies that $T = N$, $S = \{1\}$, and $N \cap \mathcal{N}_G(X) = \{1\}$. Consequently, $G = [N]\mathcal{N}_G(X)$.

Let D be another complement to N in G. Then, by Lemma 1.11, $NX = [N](NX \cap D)$. By Theorem 2.13, there is an element $t \in NX$ such that $NX \cap D = t^{-1}Xt$. Given that $D \leq \mathcal{N}_G(NX \cap D)$, the subgroup $D^{t^{-1}} \leq \mathcal{N}_G(X)$. However, by Corollary 1.13, $D^{t^{-1}}$ is also a complement to N in G with $D^{t^{-1}} \leq \mathcal{N}_G(X)$. This means $D^{t^{-1}} = \mathcal{N}_G(X)$. □

G. Higman then uses the results proven by D.G. Higman [59] and Gaschütz [41] to prove the following results.

Theorem 3.25. [Theorem 1a in G. Higman [60]] *Let N be an abelian normal subgroup of a finite group G. Suppose that for each prime p, there is a p'-subgroup X of G such that NX is normal in G, but not N_0X for any normal subgroup N_0 of G properly contained in N and of p-power index in N. Then, N is complemented in G, and any two complements of N are conjugate.*

Higman's Proof. If p does not divide the order of N, then $X = \{1\}$ satisfies the condition of the theorem. If p does divide the order of N, then $N = P \times Q$, where P is a p-group and Q is a p'-group. By Theorem 3.24, N/Q has a complement in G/Q and any two complements of N/Q are conjugate in G/Q. Thus, by Theorem 4.2, G splits over N. Then, by Theorem 3.23, all complements of N in G are conjugate. □

Theorem 3.26. [Theorem 2 in G. Higman [60]] *If G and N satisfy the hypothesis of either Theorem 3.24 or Theorem 3.25, then N is complemented in any group H which contains both G and N as normal subgroups and any two complements are conjugate.*

Higman's Proof. Suppose the conditions of Theorem 3.24 are satisfied for some subgroup X. Let M/N be the largest normal p'-subgroup of G/N. By Theorem 2.12, $M = [N]Y$ for some subgroup Y of M. Since $Y \cong M/N$ and $(|Y|, |N|) = 1$, $X \leq Y$. The conditions of Theorem 3.24 are satisfied replacing X with Y. Clearly, $NY = M \lhd G$. If there is a normal subgroup N_0 of G such that $N_0 < N$ and $N_0Y \lhd G$ (and M), then $N_0X \cap NX \lhd G$ (and M). If H is any group such that $G \trianglelefteq H$ and $N \trianglelefteq H$, then NY/N is characteristic in G/N and thus $NY \lhd H$. However, by the hypothesis, N_0Y is not normal in G. Thus, N has a complement in H and any two complements are conjugate.

A similar argument show the same result if the conditions of Theorem 3.25 are satisfied. □

Baer [2] returns to this topic in 1957. One of the avenues of his investigation was looking at the structure and properties of minimal normal and maximal subgroups. The results listed here are not deep results, but they are contributions to the theme.

The following lemma is motivated by Baer's interest in the structure of groups G having two minimal normal subgroups and a maximal subgroup with a trivial core. Recall that for a subgroup H of a group G, the **core of H**, denoted by H_G or $\mathrm{Core}_G(H)$, is the largest normal subgroup of G contained in H. Equivalently, the core of H in G can be defined as $\mathrm{Core}_G(H) = \cap_{g \in G} g^{-1}Hg$.

Lemma 3.27. [Lemma 2.2 in Baer [2]] *If S is a maximal subgroup of the group G, if $S_G = \{1\}$, and if $N \neq \{1\}$ is a normal subgroup of G, then $C_G(N) \cap S = \{1\}$ and $C_G(N)$ is either $\{1\}$ or a minimal normal subgroup of G.*

Baer's Proof. Given that $N \nleq S$ and S is maximal in G, $G = NS$. Furthermore, since $\mathcal{C}_G(N) \lhd G$, $\mathcal{C}_N(N) \cap S \lhd S$ and $S \leq \mathcal{N}_G(\mathcal{C}_G(N) \cap S)$. Given that N commutes elementwise with every element in $\mathcal{C}_G(N)$, $\mathcal{N}_G(\mathcal{C}_G(N) \cap S) = G$ and $\mathcal{C}_G(N) \cap S \lhd G$. This implies $\mathcal{C}_G(N) \cap S = \{1\}$.

Suppose there exists a nontrivial normal subgroup M of G such that $M < \mathcal{C}_G(N)$. Given that $M \nleq S$, $G = MS$. By the modular identity (Lemma 1.6),

$$\mathcal{C}_G(N) = G \cap \mathcal{C}_G(N) = MS \cap \mathcal{C}_G(N) = M(\mathcal{C}_G(N) \cap S) = M.$$

This means $\mathcal{C}_G(N) = \{1\}$ or $\mathcal{C}_G(N)$ is a minimal normal subgroup of G □

The corollary to Lemma 3.27 is of interest here.

Corollary 3.28. [Corollary 2.2 in Baer [2]] *If S is a maximal subgroup of a group G, if $S_G = \{1\}$, and if A and B are two different minimal normal subgroups of G, then*
(i) $G = AS = BS$ *and* $A \cap S = B \cap S = \{1\}$,
(ii) *A is the centralizer of B in G, and*
(iii) *A, B, and $AB \cap S$ are isomorphic non-abelian groups.*

The final result of this section is due to Bercov. It was well known that if N is a minimal normal subgroup of a group G, then $N \cong H_1 \times \cdots \times H_n$, where, for $1 \leq i, j \leq n$, $H_i \cong H_j \cong H$ for some simple group H. In his paper, Bercov [15] considers the case that H is nonabelian and that the $Inn(H)$ has a complement in $Aut(H)$. This does occur. For example, for $n \geq 5$, with $n \neq 6$, $Aut(\mathcal{A}_n)$ splits over $Inn(\mathcal{A}_n) \cong \mathcal{A}_n$ since $Aut(\mathcal{A}_n) \cong \mathcal{S}_n$. A complete classification of those nonabelian simple groups G which split over $Inn(G)$ was given by Lucchini, Menegazzo, and Morgi in [79].

Assume that H is nonabelian and that $Aut(H) = [Inn(H)]C(H)$, where $C(H) \leq Aut(H)$. For any $x \in G$, H^x is a direct factor in N. The subgroup N is the direct product of H^{t_i} with each t_i a coset representative of $\mathcal{N}_G(H)$. Let τ denote the permutation of the direct factors of N induced conjugation by x. Thus, $H^{t_i x} = H^{t_{\tau(i)}}$ or $H = H^{t_i x t_{\tau(i)}^{-1}}$. Let $\alpha_i(x)$ be the automorphism of H induced by $t_i x t_{\tau(i)}^{-1}$, and let $C = \{x \in G \mid \alpha_i(x) \in C(H) \text{ for all } i\}$. Now Bercov's result can be stated.

Theorem 3.29. [Theorem 1 in Bercov [15]] *The subgroup C is a complement for N in G.*

4 Reduction theorems

While Ore proved the first splitting reduction theorem in 1939 (Theorem 3.5), Gaschütz [41] was the first person to complete an in-depth investigation of this area. His work set the benchmark, and he proved three big reduction results that are still the standard today. Any reduction theorem proven after 1952 has its origin in Gaschütz's work.

Before getting to Gaschütz's results, a revised version of Ore's 1939 result (Theorem 3.5) is stated first.

Theorem 4.1. [Ore [86]] *Let B be a normal subgroup of a group G. If there exists a normal subgroup A of G such that $B \leq A$, $A = [B]H$ with $\mathcal{N}_A(H) = H$, and all complements of B in A are conjugate. Then, $G = [B]K$, where $K = \mathcal{N}_G(H)$, $\mathcal{N}_G(K) = K$, and all complements of B in G are conjugate.*

While Ore begins this chapter, it is Gaschütz who really gets it going. In 1952, he published a major paper [41] on normal subgroup complementation, concentrating on abelian normal subgroups (a number of results from this paper have already been presented in Chapter 3 and will appear in later chapters). He was motivated by work done in the area of splitting by Schur, Ore, Zassenhaus, and Bergström. By constructing what he called reduced factor systems, he was able to prove his reduction theorems. His greatest insight was to reduce the question of a group splitting over an abelian normal subgroup to a splitting question involving Sylow subgroups.

Before stating the first of Gaschütz's reduction results, recall Lemma 1.11 that states that if $G = [N]H$ and $M \trianglelefteq G$ with $M \leq N$, then $G/M = [N/M]HM/M$. Gaschütz was able to employ Sylow subgroups to, under certain conditions, go the other direction.

Theorem 4.2. [1. Reduktionssatz in Gaschütz [41]] *If A is an abelian normal subgroup in a group G, then $A = A_{p_1} \times \cdots \times A_{p_t}$ with each A_{p_i}, for $1 \leq i \leq t$, the Sylow p_i-subgroup of A with $A_{p_i} \trianglelefteq G$. Let $A^*_{p_r} = A_{p_1} \times \cdots \times A_{p_{r-1}} \times A_{p_{r+1}} \times \cdots \times A_{p_t}$. Then, G splits over A if and only if $G/A^*_{p_r}$ splits over $A/A^*_{p_r}$ for each r, where $1 \leq r \leq t$.*

Beweis. ($\Rightarrow$) Suppose $G = [A]K$ for $K \leq G$. Then, by Lemma 1.11, $G/A^*_{p_r} = [A/A^*_{p_r}]KA^*_{p_r}/A^*_{p_r}$ for each r, where $1 \leq r \leq t$.

($\Leftarrow$) If $t = 1$, then the result trivially follows. Assume $t \geq 2$ and proceed by induction of t. Consider the prime p_1. Then, $A = A_{p_1} \times A^*_{p_1}$, where there exists at least one prime q that divides that order of $A^*_{p_1}$. Consequently, $A^*_{p_1} = A_q \times A^*_q$.

Given that $G/A^*_{p_1}$ splits over $A/A^*_{p_1}$, there exists a subgroup H of G such that $G/A^*_{p_1} = [A/A^*_{p_1}]H/A^*_{p_1}$. Since $A = A_{p_1} \times A^*_{p_1}$ with $A^*_{p_1} \lhd G$, it follows by Lemma 1.11 that $G = [A_{p_1}]H$ with $A \cap H = A^*_{p_1}$. By a similar argument, $G = [A_q]K$ with $K \leq G$ and $A_{p_1}A^*_q \leq K$.

DOI 10.1515/9783110480214-004

Now consider the subgroup $A^*_{p_1}$. Clearly, $A^*_{p_1}$ is abelian and normal in G. Since $G = [A_q]K$, it follows from Lemma 1.11 that $G/A^*_q = [A_qA^*_q/A^*_q]K/A^*_q$. Thus, for each prime q that divides the order of $A^*_{p_1}$, G/A^*_q splits over $A^*_{p_1}/A^*_q$. The induction hypothesis implies $G = [A^*_{p_1}]L$ for $L \leq G$.

A simple calculation using the modular identity (Lemma 1.6) shows

$$H = G \cap H = A^*_{p_1} L \cap H = A^*_{p_1}(L \cap H)$$

and

$$G = A_{p_1}H = A_{p_1}A^*_{p_1}(L \cap H) = A(L \cap H).$$

Furthermore, given that

$$A \cap (L \cap H) = A \cap H \cap L = A^*_{p_1} \cap L = \{1\},$$

it follows that $G = [A](L \cap H)$ and G splits over A. □

It can be shown that all the complements of A in G are conjugate if and only if all of the complements of A/A'_{p_r} in G/A'_{p_r} are conjugate (see page 244 in [103]). The next result is the most famous reduction theorem by Gaschütz.

Theorem 4.3. [2. Reduktionssatz in Gaschütz [41]] *Let A be an abelian normal subgroup of a group G with $|A| = p^\alpha$. The group G splits over A if and only if a Sylow p-subgroup of G splits over A.*

The reader may not be familiar with this form of Gaschütz's Reduktionssatz (Theorem 4.3) as it is usually stated in the following, but equivalent, form.

Theorem 4.4. [Gaschütz] *Let A be an abelian normal subgroup of a group G. The group G splits over A if and only if for each prime p dividing $|A|$, each Sylow p-subgroup P of G splits over $P \cap A$.*

Before proving Theorem 4.4, a proof that Theorem 4.3 and Theorem 4.4 are equivalent is given.

Proof of Equivalence. Gaschütz's Theorem 4.3 clearly follows from Theorem 4.4. Proceed by assuming as given Theorem 4.3. Let A be an abelian normal subgroup of G. Then, $A = A_{p_1} \times \cdots \times A_{p_r}$, where each A_{p_i} is the Sylow p_i-subgroup of A with $A_{p_i} \trianglelefteq G$.

First, suppose that G splits over A or that $G = [A]H$ for some subgroup H of G. Without loss of generality, consider A_{p_1}. Since $G = (A_{p_1} \times A_{p_2} \times \cdots \times A_{p_r})H$ with $A_{p_2} \times \cdots \times A_{p_r} \triangleleft G$, it follows from Lemma 1.11 that $G = [A_{p_1}](A_{p_2} \times \cdots \times A_{p_r}H)$. By Theorem 4.3 there is a Sylow p_1-subgroup P_1 of G such that P_1 splits over A_{p_1}. Given that $P_1 \cap A = A_{p_1}$, P_1 splits over $P_1 \cap A$.

For the reverse direction, it follows that Theorem 4.4 holds if A is a p-group. Assume otherwise and proceed by induction on r. To simplify the notation, let $A_{p_1} = A_p$ and $Q = A_{p_2} \times \cdots \times A_{p_r}$.

Since a Sylow p-subgroup splits over A_p, Theorem 4.3 indicates that the group G splits over A_p and $G = [A_p]H$ for some subgroup H of G. Thus, by the modular identity (Lemma 1.6) $A = G \cap A = A_pH \cap A = A_p(H \cap A)$. Since $A_p \cap (H \cap A) = (A_p \cap H) \cap A = \{1\} \cap A = \{1\}$, it follows that $H \cap A = Q$. By the induction, $G = [Q]K$ for some subgroup K of G. Thus, again by the modular identity (Lemma 1.6)

$$H = G \cap H = QK \cap H = Q(K \cap H)$$

and

$$G = A_pH = A_pQ(K \cap H) = A(K \cap H).$$

Since $A \cap K \cap H = A \cap H \cap K = Q \cap K = \{1\}$, $G = [A](K \cap H)$ and G splits over A. □

Proof of Theorem 4.4. (⇒) For each prime p that divides the order of A, $A = A_p \times A_p^*$, where A_p is the Sylow p-subgroup of A, $(|A_p|, |A_p^*|) = 1$ and both A_p and A_p^* are normal in G. Since G splits over A, $G = [A]H$ for $H \leq G$. By Lemma 1.11, $G = [A_p]A_p^*H$.

Now let P be a Sylow p-subgroup of G. Then, $A_p \leq P$ and by the modular identity (Lemma 1.6) $P = G \cap P = A_p(A_p^*H) \cap P = A_p(A_p^*H \cap P)$. Given that $A_p \cap A_p^*H \cap P = A_p \cap A_p^*H = \{1\}$, it follows that $P = [A_p](A_p^*H \cap P)$ and that P splits over $A_p = P \cap A$.

(⇐) If A is a p-group, then the result follows from Theorem 1.80. Now suppose that two or more primes divide the order of A and proceed by induction on the order of G. Then, $A = A_p \times A_p^*$, where p is a prime dividing the order of A and A_p is the Sylow p-subgroup of A with $(|A_p|, |A_p^*|) = 1$. Note that both A_p and A_p^* are normal in G.

Consider G/A_p^*, and let P/A_p^* be a Sylow p-subgroup of G/A_p^*. Since A_p^* is an abelian normal subgroup of P with $(|A_p^*|, [P : A_p^*]) = 1$, then by Theorem 2.12 $P = [A_p^*]L$ for $L \leq P$. Given that L is a Sylow p-subgroup of G, $L = [A_p]K$, for $K \leq L$, which implies that $P = [A_p^*A_p]K$ or that $P = [A]K$. By Lemma 1.11, it follows that $P/A_p^* = [A/A_p^*]KA_p^*/A_p^*$ or that P/A_p^* splits over A/A_p^*. By the induction hypothesis, $G/A_p^* = [A/A_p^*]H/A_p^*$ or $G = AH$ with $A \cap H = A_p^*$.

Given that the order of $[G : H]$ is some power of p, every Sylow q-subgroup of H is a Sylow q-subgroup of G for primes $q \neq p$. Thus, by our induction hypothesis, $H = [A_p^*]L$ for $L \leq K$. Given that $G = AH = AA_p^*L = AL$ and that $A \cap L \subseteq A_p^* \cap L = \{1\}$, it follows that G splits over A. □

Theorem 4.3 by Gaschütz is not true if A is not abelian. This is demonstrated in the following counterexample due to Baer.

Example 4.5. First, consider the group $L = SL(2,3)$. Then, L is generated by the elements $a = \begin{pmatrix} 1 & 1 \\ 1 & -1 \end{pmatrix}$, $b = \begin{pmatrix} 0 & -1 \\ 1 & 0 \end{pmatrix}$, and $c = \begin{pmatrix} -1 & 1 \\ -1 & 0 \end{pmatrix}$ such that $a^4 = b^4 = c^3 = 1$ with $a^2 = b^2$ and the relations $c^{-1}ac = b$, $c^{-1}bc = ab$, $bab = a$, and $b^{-1}ab = a^{-1}$. Then, $Q = \langle a, b\rangle$ is normal in L, with Q isomorphic to the quaternion group of order 8, where the center $Z(L) = \langle a^2\rangle$. Let $D = \langle d\rangle$ be a cyclic group of order 4. Construct the group $G = LD$ of order $2^4 3$, where each element of L commutes with each element of D and $a^2 = d^2$.

The subgroup $P_2 = DQ$ is a Sylow 2-subgroup of G of order 16. Let $E = \langle ad\rangle$. Since $(ad)^2 = a^2d^2 = 1$, E is of order 2 such that $E \cap Q = \{1\}$. Thus, $P_2 = [Q]E$. Thus, for each Sylow 2-subgroup S_2 of G, S_2 will split over $S_2 \cap Q$. And, trivially, each Sylow 3-subgroup S_3 will also split over $S_3 \cap Q$. However, as will be shown, the group G does not split over Q.

Suppose G splits over Q. Since G/Q is cyclic of order 6, there must be some complement of Q in G that contains c. So let $G = [Q]H$, where $c \in H$. Since $P_2 = DQ$ is Sylow 2-subgroup of G and G is solvable, $G = P_2C$, where $C = \langle c\rangle$. By the modular identity (Lemma 1.6) $H = G \cap H = P_2C \cap H = C(P_2 \cap H)$. Since $|C| = 3$, this implies that $|P_2 \cap H| = 2$. Furthermore, given that G/Q is cyclic, $P_2 \cap H \le C_G(c)$. However, $C_{P_2}(c) = \langle d\rangle$. Given that $|P_2 \cap H| = 2$, we get $P_2 \cap H = \langle d^2\rangle = \langle a^2\rangle = Z(L)$. Thus, $Z(L) = P_2 \cap H \le Q \cap H = \{1\}$, which is a contradiction. Consequently, each Sylow subgroup of G splits over Q, but G does not.

The final reduction theorem developed by Gaschütz is now stated.

Theorem 4.6. [Satz 1 in Gaschütz [41]] *Let A be an abelian normal subgroup of a group G and let U be a subgroup of G such that $A \le U \le G$. If $([G : U], |A|) = 1$, then G splits over A if and only if U splits over A.*

Beweis. Suppose $G = [A]H$ for some subgroup H of G. Then, by Lemma 1.11 $U = [A](H \cap U)$ and U splits over A. The converse follows by Theorem 1.80. □

Gaschütz makes two comments in his paper concerning Theorem 4.6. First, he notes that if $U = A$, then "den Satz von Schur," or Theorem 2.12 with N abelian, follows directly from his result (A trivially splits over A). Second, he notes, without proof, that under certain conditions, all complements of A in G will be conjugate. In his book *Endliche Gruppen I*, Huppert restates Theorem 4.6 by Gaschütz adding the conjugacy of complements, which follows from Theorem 1.80.

Theorem 4.7. [Hauptsatz I.17.4 in Huppert [66]] *Let A be an abelian normal subgroup of a group G, and let U be a subgroup of G such that $A \le U \le G$ with $[G : U] = k$. Also suppose that the mapping $a \to a^k$ (for $a \in A$) has an inverse τ.*

(i) *If A has a complement in U, then A has a complement in G.*
(ii) *If A has a complement in U and if all complements of A in U are conjugate, then all complements of A in G are conjugate.*

Now to a result by D.G. Higman who was motivated by the Schur-Zassenhaus theorem and the results by Gaschütz [41]. As noted in Example 4.5, Theorem 4.3 is not true if the normal subgroup is not abelian. D.G. Higman's insight was to develop a reduction condition for splitting involving Sylow subgroups for the case where the normal subgroup is not abelian.

Theorem 4.8. [Theorem 1 in D.G. Higman [59]] *For a subgroup C and a normal subgroup N of a group G, then $G = [N]C$ if and only if G is the reduced product of N by C and there exists, for each prime p dividing $|G|$, a Sylow p-subgroup P of G and a complement of $N \cap P$ in P which is part of C.*

Essentially, D.G. Higman is saying that $G = [N]C$ if and only if $G = NC$ is a reduced product and $P = [N \cap P](C \cap P)$.

D.G. Higman's Proof. ($\Rightarrow$) If $G = [N]C$, then clearly G is the reduced product of N with C. Let P be a Sylow p-subgroup of G such that P contains a Sylow p-subgroup P_c of C. Thus, $N \cap P$ is a Sylow p-subgroup of N with $N \cap P \cap P_c = \{1\}$. By the modular identity (Lemma 1.6), $P_c(N \cap P) = NP_c \cap P = P$, and it follows that $P = [N \cap P](P \cap C)$.

($\Leftarrow$) Suppose $G = NC$ is a reduced product. Then, by Lemma 1.60, $N \cap C \leq \Phi(C)$, and by Lemma 1.51, $N \cap C$ is nilpotent. Let p divide the order of G. Then, there exists a Sylow p-subgroup P of G such that $P = [N \cap P]P^*$ for $P^* \leq C$. Thus, $P^* = P \cap C$ and $P = [N \cap P](P \cap C)$.

Let P_c be a Sylow p-subgroup of C such that $P \cap C \leq P_c$, and let S_p be a Sylow p-subgroup of G such that $P_c \leq S_p$. Thus, there exists an element $g \in G$ such that $P^g = S_p$. Therefore,

$$S_p = P^g = (N \cap P)^g (P \cap C)^g = (N \cap S_p)(P \cap C)^g.$$

Given that $(N \cap S_p) \cap (P \cap C)^g = \{1\}$, $S_p = [N \cap S_p](P \cap C)^g$.

Now suppose there is a nontrivial element x in $(N \cap S_p) \cap (P \cap C)$. This implies $x^g \in (P \cap C)^g$. Since $x \in P$ and $x \in N$, it follows that $x^g \in P^g \cap N = S_p \cap N$. This implies that $(N \cap C) \cap (P \cap C)^g \neq \{1\}$, a contradiction. This means $(N \cap S_p) \cap (P \cap C) = \{1\}$. Given that $|P \cap C| = |(P \cap C)^g|$, it follows that $S_p = [N \cap S_p](P \cap C)$. Consequently, using the modular identity (Lemma 1.6)

$$\begin{aligned} P_c &= P_c \cap S_p \\ &= P_c \cap (N \cap S_p)(P \cap C) \\ &= (P \cap C)(P_c \cap N \cap S_p) \\ &= (P \cap C)(N \cap P_c) \\ &= (N \cap P_c)(P \cap C) \\ &= (N \cap C \cap P_c)(P \cap C), \end{aligned}$$

which implies $P_c = [N \cap C \cap P_c](P \cap C)$.

Note that $N \cap C \cap P_c$ is the Sylow p-subgroup of $N \cap C$, which is nilpotent. Given that $N \cap C \lhd C$, the Sylow p-subgroup of $N \cap C$ is normal in C and contained in each Sylow p-subgroup of C. Thus, for each prime p dividing the order of C, each Sylow p-subgroup of C splits over the p-component of $N \cap C$.

If $A = N \cap C$ is abelian, C splits over $N \cap C$ by Theorem 4.4. This is a contradiction as $N \cap C \leq \Phi(C)$. Thus, A is not abelian. Consider the commutator subgroup A' of A and the group C/A'. The subgroup P_cA'/A' is a Sylow p-subgroup of C/A'. Recall that $P_c = [N \cap C \cap P_c](P \cap C) = [A \cap P_c](P \cap C)$, which implies that $A \cap P_c \cap P \cap C = A \cap P \cap C = \{1\}$. Given that (using the modular identity (Lemma 1.6))

$$\begin{aligned} A'(A \cap P_c) \cap A'(P \cap C) &= A'P_c \cap A \cap A'(P \cap C) \\ &= A'P_c \cap A'(P \cap C) \cap A \\ &= A'(P \cap C) \cap A \\ &= A'(P \cap C \cap A) \\ &= A', \end{aligned}$$

$P_cA'/A' = [A'(A \cap P_c)/A']A'(P \cap C)/A'$. Applying Theorem 4.4 to C/A', it follows that $C/A' = [A/A']D/A'$. This implies that $C = AD$ with D a proper subgroup of C. This is again a contradiction as $A = N \cap C \leq \Phi(C)$. Thus, $A = N \cap C = \{1\}$ and $G = [N]C$. □

This chapter ends with a few results by Šemetkov. He was motivated by Wielandt's [109] address to the 1958 International Congress of Mathematicians held in Edinburgh. In his address, Wielandt gave an up-to-date overview of what he called the "complement problem." He mentions Zassenhaus's result (Theorem 2.13) and states how many, including Zassenhaus, believed that the conditions on the normal subgroup or its quotient group were not needed for all complements to be conjugate. Wielandt also mentions Gaschütz's Theorem 4.3 and that it would be better the remove the requirement that the normal subgroup be abelian. As shown by Example 4.5, if the normal subgroup is not abelian, then the result does not hold. While not mentioned by Wielandt, D.G. Higman in 1954 had proven a result (Theorem 4.8) that did not require the normal subgroup to be abelian. In 1970, Šemetkov was also able to able to improve upon Gaschütz's result.

In his paper, Šemetkov [99] introduces the concept of a π-complement. Let π be a set of primes. A subgroup H of G has a **π-complement** in G if there exists a subgroup K of G such that $G = HK$ and $|H \cap K|$ is not divisible by any of the primes in π. The following theorem is at the foundation of his complementation results from [99].

Theorem 4.9. [Theorem 1 in Šemetkov [99]] *A normal subgroup N of a group G has a π-complement in G if $S_p \cap N$ is abelian and complemented in S_p, where S_p is a Sylow p-subgroup of G, for each p in π.*

Šemetkov's Proof. Let a group G be a counterexample of minimal order. Thus, there exists a nonempty set of primes π and a subgroup N of G, where $N \lhd G$ and N has no π-complement in G. Note that if $\pi = \emptyset$, then G is the π-complement to N in G. Furthermore, if no prime p in π divides the order of N, then G is again a π-complement to N in G. Thus, there is at least one prime in π that divides the order of N.

Let ρ be a subset of π such that N has no ρ-complement in G and N has a ρ_1-complement in G for every proper subset ρ_1 of ρ that does not contain a divisor of $|N|$. By the work done in the first paragraph, it follows that ρ is nonempty. Let $p \in \rho$ such that p divides the order of N, and let $\sigma = \rho - \{p\}$. Then, N has a σ-complement H in G.

Since $G = NH$, there is a subgroup K of G such that $K \leq H$ and $G = NK$ is a reduced product of N with K, where $N \cap K \leq \Phi(K)$ (see Lemma 1.60). Note that $N \cap K$ is nilpotent (Lemma 1.51) with $N \cap K \leq N \cap H$. Thus, no primes in σ divide the order of $N \cap K$. Thus, K is a σ-complement of N in G. In addition, since N has a ρ-complement in G, p divides $|N \cap K|$.

Let S_p^* be a Sylow p-subgroup of K. Then, $S_p^* \leq S_p$, a Sylow p-subgroup of G. Let $P = S_p \cap N$, which is a Sylow p-subgroup of N. Given that p does not divide $[G : S_p^*P]$, the modular identity (Lemma 1.6) implies

$$S_p^*P = S_p^*(N \cap S_p) = S_p^*N \cap S_p = S_p$$

or $S_p = S_p^*P$. It follows that

$$P_1 = S_p^* \cap (N \cap K) = S_p^* \cap N = S_p^* \cap S_p \cap N = S_p^* \cap P$$

is a nontrivial Sylow p-subgroup of $N \cap K$ (recall that p divides the order of $N \cap K$). Now consider $\mathcal{N}_G(P_1)$.

Case 1: Suppose $\mathcal{N}_G(P_1) \neq G$.

Since $N \cap K$ is nilpotent with $N \cap K \lhd K$, it follows that $K \leq \mathcal{N}_G(P_1)$. Given that $P = S_p \cap N$ is abelian, $P \leq \mathcal{N}_G(P_1)$. This implies that $S_p = S_p^*P \leq \mathcal{N}_G(P_1)$. By the hypothesis, P has a complement in S_p.

Let $q \in \rho$ such that $q \neq p$, Let S_q^* be a Sylow q-subgroup of K. Then, $S_q^* \leq S_q$, a Sylow q-subgroup of $\mathcal{N}_G(P_1)$. Let $Q = S_q \cap \mathcal{N}_G(P_1) \cap N = S_q \cap N$ be a Sylow q-subgroup of $\mathcal{N}_G(P_1) \cap N$. It follows directly using the modular identity (Lemma 1.6) that

$$S_q^*Q = S_q^*(S_q \cap N) = S_q^*N \cap S_q = S_q.$$

Note that $S_q^* \cap Q \leq N \cap K$, where $|N \cap K|$ is not divisible by the primes in σ. Thus, q does not divide the order of $N \cap K$ and $S_q^* \cap Q = \{1\}$. This means $\mathcal{N}_G(P_1)$, $\mathcal{N}_G(P_1) \cap N$, and σ satisfy the hypothesis of the theorem. This implies $\mathcal{N}_G(P_1) \cap N$ has a σ-complement

J in $\mathcal{N}_G(P_1)$ ($\mathcal{N}_G(P_1) = (\mathcal{N}_G(P_1) \cap N)J$). Given that $K \le \mathcal{N}_G(P_1)$ and $G = NK$, it follows using the modular identity (Lemma 1.6) that

$$\begin{aligned} G &= NK \\ &= N\mathcal{N}_G(P_1) \\ &= N(J(\mathcal{N}_G(P_1) \cap N)) \\ &= N(JN \cap \mathcal{N}_G(P_1)) \\ &= N\mathcal{N}_G(P_1) \cap JN \\ &= G \cap JN \\ &= NJ. \end{aligned}$$

Since $N \cap J \le N \cap \mathcal{N}_G(P_1)$, the order of $N \cap J$ is not divisible by any primes in σ. Thus, J is a π-complement to N in G, a contradiction.

Case 2: $\mathcal{N}_G(P_1) = G$.

It follows that $P_1 \lhd G$. Let C be the product of all of the normal subgroups of N that do not have composition factors of order p. Thus, C is a characteristic subgroup of N with no composition factors of order p. Given that the Sylow p-subgroup P of N is abelian, Satz 3.3 in [65] implies there exists a normal subgroup N_1 of N such that N/N_1 is p-solvable and every abelian composition factor of N_1 has order relatively prime to p. Thus, $N_1 \le C$ with $N/C \cong (N/N_1)/(C/N_1)$, which implies that N/C is p-solvable. By Hauptsatz VI.6.6 in [66] (originally proven in [57]), PC/C is a Sylow p-subgroup of N/C and $PC/C \lhd N/C$. By the hypothesis, $S_p = [P]P^*$ and $S_p \cap C = S_p \cap N \cap C = P \cap C$ is a Sylow p-subgroup of C. Given that $C \le N$, it follows that $P^* \cap C = \{1\}$. Thus, $PC/C \cap P^*C/C = \{1_{N/C}\}$ and $S_pC/C = [PC/C]P^*C/C$. By Theorem 4.6, it then follows that $KPC/C = [PC/C]U/C$.

Let $q \in \rho$ such that $q \ne p$. Since q does not divide $|K \cap PC|$, there exists a Sylow q-subgroup S_q of MPR such that $S_q = Q^*Q$, where Q is a Sylow q-subgroup of M and Q^* is a Sylow q-subgroup of C with $Q^* \cap Q = \{1\}$. Given the fact that $KPC/C = [PC/C]U/C$, the Sylow q-subgroup of KPC and U have the same order. Thus, there exists an element $x \in KPC$ such that $S_q^x \le U$. By the hypothesis, any Sylow q-subgroup of C has a complement in a Sylow q-subgroup of U which contains it. Given that $O^p(C) = C$, Theorem 6.2 implies that a Sylow p-subgroup of C has a complement in a Sylow p-subgroup of U. Thus, the hypothesis of the theorem is satisfies by ρ, U, and C. Given that $|U| < |G|$, the result holds for U. This implies that C has a ρ-complement L in U or that $U = [C]L$. Given that $UPC = KPC$ with $PC \le N$, it follows that, since $G = NK$, $G = NL$. In addition, for each $q \in \rho$, each Sylow q-subgroup of L has the same order as a Sylow q-subgroup of G/N. Thus, $|N \cap L|$ is a ρ'-number, which is a contradiction. □

The following result directly follows Theorem 4.9 and is a generalization of Gaschütz's Theorem 4.4.

Theorem 4.10. [Theorem 2 in Šemetkov [99]] *A normal subgroup N of a group G is complemented in G if $S_p \cap N$ is abelian and complemented in S_p for each prime divisor of $[G : N]$ and S_p a Sylow p-subgroup of G.*

Beweis. Let π denote the set of primes diving $[G : N]$. By Theorem 4.9, N has a π-complement K in G, where the order of $N \cap K$ is a π'-number. Given that $|NK/N| = |K/(N{\cap}K)|$, $N{\cap}K$ is normal in K and relatively prime to its index. Thus, by Theorem 2.12, $K = [N \cap K]L$ for $L \leq K$. Thus, by Lemma 1.11, $G = [N]L$ and N has a complement in G. □

In addition, the following generalization of Gaschütz's Theorem 4.4 was also proven by Šemetkov.

Theorem 4.11. [Theorem 3 in Šemetkov [99]] *A normal subgroup N of a group G is complemented in G if, for each prime p, a Sylow p-subgroup of N is abelian and complemented in any Sylow p-subgroup of G containing it.*

This chapter ends with a simple result from Christensen [28].

Theorem 4.12. [Variation of Theorem 5.3 in Christensen [28]] *Let N be a normal subgroup of a solvable group G such that N is complemented in G. Then, the intersection of N with any Sylow p-subgroup S_p of G is complemented in S_p for all primes p.*

Christensen's Proof. Let $|G| = p_1^{r_1} \cdots p_t^{r_t}$. By Theorem 1.35 there are Sylow subgroups S_{p_i}, for $1 \leq i \leq t$, of G such that $G = S_{p_1} \cdots S_{p_t}$. By Lemma 1.36, $N = (N \cap S_{p_1}) \cdots (N \cap S_{p_t})$.

Since G splits over N, $G = [N]H$ for $H \leq G$. Given that H is solvable, $H = T_{p_1} \cdots T_{p_t}$ for Sylow subgroups T_{p_j}, for $1 \leq j \leq t$, of H with the possibility that $T_{p_j} = \{1\}$. For each T_{p_j}, there exists an element $g_j \in G$ such that $T_{p_j}^{g_j} \leq S_{p_j}$. However, given that $G = [N]H$, this implies $S_{p_j} = (N \cap S_{p_j})T_{p_j}^{g_j}$ as $T_{p_j}^{g_j} \leq H^{g_j}$, where H^{g_j}, by Corollary 1.13 is also a complement of N in G. Given that $N \cap S_{p_j} \trianglelefteq S_{p_j}$ and $N \cap S_{p_j} \cap T_{p_j}^{g_j} = \{1\}$, $S_{p_j} = [N \cap S_{p_j}]T_{p_j}^{g_j}$. □

5 Subgroups in the chief series, derived series, and lower nilpotent series

This chapter begins with a two simple observations.

Lemma 5.1. *Let G be a group, let N be a normal subgroup of G, let $\{1\} = N_0 \lhd N_1 \lhd \cdots \lhd N_t = N$ be part of a chief series for G containing N, and let H be a subgroup of G that avoids each chief factor N_i/N_{i-1} for $1 \leq i \leq t$.*
(i) $N \cap H = \{1\}$.
(ii) *If $G = NH$, then $G = [N]H$.*

Beweis. To prove (i), first note that $N_t \cap H \leq N_{t-1}$ as H avoids N_t/N_{t-1}. If $t = 1$, then $N \cap H = \{1\}$. Now assume $t \geq 2$ and proceed by induction on t, assuming that $N \cap H \leq N_1$. Given that N_1 is minimal normal in G, it is avoided by H. Consequently, $N_1 \cap H = \{1\}$. Given that $N \cap H \leq N_1$, it follows that

$$N \cap H \leq N_1 \cap H = \{1\}.$$

Since $N \cap H = \{1\}$, (ii) directly follows. □

Lemma 5.2. *Let G be a group, let N be a normal subgroup of G, and let $\{1\} = N_0 \lhd N_1 \lhd \cdots \lhd N_t = N$ be part of a chief series for G containing N. If each chief factor N_i/N_{i-1}, for $1 \leq i \leq t$, is a complemented chief factor, then G splits over N.*

Beweis. Given that N_1 is complemented, $G = [N_1]H_1$ for some maximal subgroup H_1 of G. If $t = 1$, the result follows. Assume $t \geq 2$ and proceed by induction assuming that $G = [N_{t-1}]H_{t-1}$ for $H_{t-1} \leq G$.

Given that N_t/N_{t-1} is a complemented chief factor, there exists a subgroup H_t of G such that $G/N_{t-1} = [N_t/N_{t-1}]H_t/N_{t-1}$. Let $g \in G$. Then, $g = n_t h_t$ for $n_t \in N_t$ and $h_t \in H_t$. Now $h_t = n_{t-1}h_{t-1}$ for $n_{t-1} \in N_{t-1}$ and $h_{t-1} \in H_{t-1}$. This implies $h_{t-1} = h_t n_{t-1}^{-1}$. Given that $N_{t-1} \leq H_t$, it follows that $h_{t-1} \in H_{t-1} \cap H_t$. Furthermore,

$$g = n_t h_t = n_t n_{t-1} h_{t-1} = n h_{t-1},$$

where $n = n_t n_{t-1} \in N_t$, resulting in $G = N_t(H_{t-1} \cap H_t)$. Since $N_t \cap H_t \cap H_{t-1} = N_{t-1} \cap H_{t-1} = \{1\}$, $G = [N_t](H_{t-1} \cap H_t)$ and G splits over N. □

These two lemmas yield the following result.

Theorem 5.3. *Let G be a solvable group, let N be a normal subgroup of G, and let $\{1\} = N_0 \lhd N_1 \lhd \cdots \lhd N_t = N$ be part of a chief series for G containing N. If each chief factor N_i/N_{i-1}, for $1 \leq i \leq t$, is avoided by a prefrattini subgroup W of G, then*

DOI 10.1515/9783110480214-005

(i) $N \cap W = \{1\}$ *and*
(ii) G *splits over* N.

Beweis. Item (i) follows directly from Lemma 5.1. To prove (ii), first note that by Theorem 1.66 that N_i/N_{i-1} for $1 \leq i \leq t$ are complemented chief factors. The result then follows directly from Lemma 5.2. □

In 1940, Hall [56] published a paper which presented a way to construct solvable groups without the use of factor systems. To avoid the difficulties with factor systems, his construction method was based on partial complemented extensions. An extension of a group G by a normal subgroup N is a **partial complemented extension** if there exists a subgroup D of G such that $G = ND$. If $N \cap D = \{1\}$, then the extension is a **complemented extension.** As noted earlier, complemented extensions are good, as each element $g \in G$ can be uniquely written as $g = nd$ with $n \in N$ and $d \in D$. Hall's paper focused on possible choices for N and D so that the solvable group G can be written as a partial complemented extension of N by D. Hall refers to these choices as $N(G)$ and $D(G)$ for a solvable group G. He states that it is natural to "seek solutions" where $N(G)$ is a characteristic subgroup of G and that $D(G)$ be a member of some characteristic conjugacy class of subgroups of G.

Hall's method involved the lower nilpotent series. Given a solvable group G, the lower central series

$$G = L_0 > L_1 > \cdots > L_n = \{1\}$$

will terminate at $\{1\}$ for some $n \geq 1$. Since each member of the lower central series is characteristic in G, he takes L_i, for $1 \leq i \leq n-1$, as possible choices for $N(G)$ as his goal was to construct non-nilpotent solvable groups. The next step was to come up with choices for $D(G)$. He then uses the concept of Sylow system normalizers that he had developed in [55].

Recall Lemma 1.33, which indicates that for a solvable normal subgroup N of a group G that $\mathcal{N}_G(\Sigma_N)$ is a partial complement of N in G. Hall then showed that if $N(G) = L_i$ and $D(G) = \mathcal{N}_G(\Sigma_{L_{i-1}})$ for $0 < i < n$, then G has a partially complemented extension satisfying the requirements.

For complementation, Hall turned his attention to A-groups.

Definition 5.4. A solvable group G is an **A-group** if its Sylow subgroups are abelian.

Clearly, nilpotent A-groups are abelian. Also note that a group with abelian Sylow subgroups need not be solvable as indicated by the alternating group $\mathcal{A}_5$. A couple of properties of A-groups are given here, but for more information on A-groups, see [66, 104].

Lemma 5.5. *Let G be an A-group.*

(i) *Any subgroup and homomorphic image of G is also an A-group.*
(ii) *The lower nilpotent series of G coincides with its derived series.*

Beweis. For (i), first note that subgroups and homomorphic images of G are solvable as G is solvable (Lemma 1.25). Given that any p-subgroup of G is contained in a Sylow p-subgroup of G and that any Sylow p-subgroup of the quotient group G/N, for $N \trianglelefteq G$, is the image of a Sylow p-subgroup of G, the result follows.

To prove (ii), let

$$G = G^0 > G^1 > \cdots > G^n = \{1\}$$

be the derived series and

$$G = L_0 > L_1 > \cdots > L_t = \{1\}$$

be the lower nilpotent series for G. Since L_1 is the smallest normal subgroup of G such that G/L_1 is abelian, $L_1 = G^1$. If $n = 1$, the result follows. Assume that $n \geq 2$ and proceed by induction on n, assuming that $L_{n-1} = G^{n-1}$. Since G^{n-1} is abelian (and nilpotent), this forces $L_n = G^n = \{1\}$. □

Hall proved the following result, which is an extension of Lemma 1.33.

Theorem 5.6. [Hall [56]] *If H/N is a nilpotent factor group of G with H solvable, then $G = N\mathcal{N}_G(\Sigma_H)$.*

Beweis. Let

$$N = H_0 \lhd H_1 \lhd \cdots \lhd H_n = H$$

be part of a chief series for H that contains N, and let $\mathcal{N}_H(\Sigma_H)$ be a System normalizer for H. Given that H/N is nilpotent, each chief factor H_i/H_{i-1}, for $1 \leq i \leq n$, is central in H. By Theorem 1.39, $S = \mathcal{N}_H(\Sigma_H)$ covers each of these chief factors.

If $n = 1$, then the chief factor H/N is covered by S and $NS \cap H = H$ or $H \leq NS$. Now suppose $n \geq 2$ and that $NS \cap H_{n-1} = H_{n-1}$. Since S covers chief factor H_n/H_{n-1}, it follows that $H_{n-1}S \cap H_n = H_n$. Consequently, $(NS \cap H_{n-1})S \cap H_n = H_n$ or $NS \cap H_{n-1}S \cap H_n = H_n$. This implies $NS \cap H_n = H_n$ or that $H \leq NS$. Given that $NS \leq H$, it follows that $H_n = H = NS$ or $H = N\mathcal{N}_H(\Sigma_H)$.

By Lemma 1.33, $G = H\mathcal{N}_G(\Sigma_H)$. The fact $\mathcal{N}_H(\Sigma_H) \leq \mathcal{N}_G(\Sigma_H)$ implies that $G = N\mathcal{N}_G(\Sigma_H)$. □

Theorem 5.7. [Hall [56]] *The system normalizers of an A-group are all complements of the derived subgroup.*

Beweis. Given an A-group G, its lower nilpotent series and its derived series coincide by Lemma 5.5. This means G' contains no central chief factors. Let Σ be a Sylow system for G. Since G/G' is nilpotent, it follows from Theorem 5.6 that $G = G'\mathcal{N}_G(\Sigma)$. Given that all of the chief factors of G contained in G' are eccentric, $\mathcal{N}_G(\Sigma)$ avoids them by Theorem 1.39. Thus, by Lemma 5.1, $G = [G']\mathcal{N}_G(\Sigma)$. □

The following two results by Hall follow directly from Theorem 5.7.

Theorem 5.8. [Hall [56]] *Let H be a normal subgroup of an A-group G. Then, $\mathcal{N}_G(\Sigma_H)$ is a complement to H' in G.*

Beweis. By Theorem 5.7, $H = [H']\mathcal{N}_H(\Sigma_H)$. In addition, by Theorem 5.6, it follows that $G = H'\mathcal{N}_G(\Sigma_H)$. However, given that $H' \cap \mathcal{N}_G(\Sigma_H) \leq \mathcal{N}_H(\Sigma_H)$,

$$H' \cap \mathcal{N}_G(\Sigma_H) \leq H' \cap \mathcal{N}_H(\Sigma_H) = \{1\},$$

and it follows that $G = [H']\mathcal{N}_G(\Sigma_H)$. □

The following result by Hall is a corollary to Theorem 5.8

Corollary 5.9. [Hall [56]] *Let G be an A-group with lower central series $G = L_0 > L_1 > \cdots > L_n = \{1\}$. Then, for each i, $0 \leq i \leq n-1$, $G = [L_{i+1}]\mathcal{N}_G(\Sigma_{L_i})$.*

Taunt [104] gave an example to show that the complements of L_i in an A-group G need not be conjugate. Specifically, he gave the following example of an A-group in which there is a complement to L_1 which is not a system normalizer.

Example 5.10. Consider the group G of order $2 \cdot 7 \cdot 13^3$ defined as follows:

$$\begin{aligned} G = \langle a, b, c, d, e \mid & a^2 = b^7 = e^{13} = d^{13} = e^{13} = 1, ad = da, cd = dc, \\ & a^{-1}ba = b^{-1}, a^{-1}ca = c^{-1}d^{-1}, b^{-1}cb = c^2d^3, b^{-1}db = c^3d^5, \\ & \text{and } e \text{ permutes with } a, b, c, d, \text{ and } e \rangle \end{aligned}$$

Now $G = \langle a, b, c, d\rangle \times \langle e\rangle$ with $L_1 = \langle b, c, d\rangle$ of order $7 \cdot 13^2$ and $Z(G) = \langle e\rangle$. The subgroup $X = \langle a, de\rangle$ is of order $2 \cdot 13$ and complements L_1, but it is not a system normalizer as it does not contain $\langle e\rangle$.

Taunt did show that all complements of L_1 of an A-group G are complemented under certain conditions.

Theorem 5.11. [Taunt [104]] *If G is a metabelian A-group, then any complement X of L_1 is a system normalizer and all complements of L_1 are conjugate.*

Beweis. Since L_1 is abelian, for each prime p_i dividing the order of G, $L_1 = P_1 \times \cdots \times P_r$, where each P_i is a Sylow p_i subgroup of L_1 with the possibility that $P_i = \{1\}$. Similarly, the subgroup $X = Q_1 \times \cdots \times Q_r$, where each Q_i is a Sylow p_i-subgroup of X. Since each $P_i \trianglelefteq G$, $S_i = P_i Q_i$ is a Sylow p_i-subgroup of G with $S_i S_j = S_j S_i$ for $1 \leq i, j \leq r$. The subgroups $\{S_1, \ldots, S_r\}$ will form a Sylow system for G. Since $L_1 = G'$, the system normalizer of this Sylow system is X and X is a system normalizer of G. Given that the system normalizers are conjugate, the result follows. □

He also proved the following special case.

Theorem 5.12. [Theorem 8.6 in Taunt [104]] *If G is an A-group of cube-free order, every complement of L_1 is a system normalizer, and every complement of L_2 is a relative system normalizer $\mathcal{N}_G(\Sigma_{L_1})$.*

The next set of results come from Schenkman. While not specifically mentioning any particular motivation, it appears that he was motivated, just as D.G. Higman was, to find conditions on when a normal subgroup is complemented and all of its complements are conjugate. In addition, he may have been trying generalize/extend Hall and Taunt's work. This is a natural assumption as he chooses as his normal subgroup the hypercommutator subgroup $D_\infty(G)$. Note that $G/D_\infty(G)$ is nilpotent (and thus solvable) meeting one of the requirements mentioned by Zassenhaus in Theorem 2.13 for all complements of a normal subgroup to be conjugate.

Theorem 5.13. [Theorem 1 and Theorem 2 in Schenkman [91]] *If a group G has an abelian hypercommutator $D_\infty(G)$, then G contains a subgroup X such that $G = [D_\infty(G)]X$ and all complements to $D_\infty(G)$ in G are conjugate.*

A proof for Theorem 5.13 is not given, as G. Higman [60] generalized Schenkman's result a year later.

Theorem 5.14. [Theorem 3 in G. Higman [60]] *If $L_r(G)$ is abelian for any integer r, it is complemented in G and any two complements are conjugate.*

Beweis. Note that if $L_r(G)$ is abelian, then $L_{r+1}(G) = \{1\}$ and G is solvable. Suppose that $L_1(G) = N$ is abelian. It will be shown that the conditions of Theorem 3.25 are satisfied. Let p divide $|N|$. Then, $N = P \times Q$, where P is the Sylow p-subgroup of N and Q is the p'-subgroup of N. Since G/N is nilpotent, $G/N = A/N \times B/N$, where A/N is the Sylow p-subgroup of G/N and B/N is the p'-subgroup of G/N. Clearly, $P \trianglelefteq G$ and $([B : P], |P|) = 1$. By Theorem 2.12, $B = [P]X$ with X a p'-subgroup of B and $B = PX = NX \lhd G$.

Suppose there exists a subgroup $N_0 \trianglelefteq G$ such that $N_0 < N$ and $[N : N_0] = p^\alpha$ with $N_0 X \trianglelefteq G$. Then, $N_0 X / N_0 \cong X / X \cap N_0$, which implies that $[N_0 X : N_0]$ is a p'-number. Furthermore, the modular identity (Lemma 1.6) implies $N_0 X \cap N = N_0 (X \cap N)$.Since X

is a p'-group, $X \cap N$ is a p'-group. Since $N_0 \leq N_0X$, we have $N_0 \leq N_0X \cap N$. However, since $N_0X \cap N = N_0(X \cap N)$ and $X \cap N$ is a p'-group, it follows that $N_0 = N_0X \cap N$. As a result,

$$N_0X/N_0 = N_0X/N \cap N_0X \cong NN_0X/N = NX/N = B/N.$$

Since B/N is nilpotent, the group N_0X/N_0 is nilpotent. Furthermore, since $[N_0X : N_0]$ is a p'-number, p does not divide $|N_0X/N_0|$. However, $|A/N_0|$ is divisible by p.

Thus, A/N_0 and N_0X/N_0 are both normal in G/N_0 and $A/N_0 \times N_0X/N_0$ is nilpotent with $A/N_0 \times N_0X/N_0 \leq G/N_0$, but $G = AB = ANX = AX$ and $G/N_0 = A/N_0 \times N_0X/N_0$. This contradicts $N = L_1(G)$. Thus, N_0 does not exists, and by Theorem 3.25, $L_1(G)$ has a complement and all complements are conjugate. This and Theorem 3.26 gives the result. □

The discussion now continues with solvable groups G and the lower nilpotent series. For a solvable group G, the lower nilpotent series

$$G = L_0 > L_1 > \cdots > L_n = \{1\}$$

will terminate at $\{1\}$ for some $n \geq 1$. For Schenkman (Theorem 5.13), $n = 2$ with L_1 abelian. In this case, $L_1 = D_\infty(G)$ and he shows that $G = [L_1]X = [D_\infty(G)]X$, for $X \leq G$, and all complements of L_1 are conjugate. This was generalized by G. Higman in Theorem 5.14, where he proved that if L_r is abelian for a positive integer r, then $G = [L_r]Y$, for $Y \leq G$, and all complements of L_r are conjugate. Note that since L_r is abelian, then $L_{r+1} = \{1\}$.

Carter [24] was motivated by the work done by Schenkman [91] and G. Higman [60] with the goal of "point(ing) out the connection between the theorems of Schenkman and G. Higman and the theory of system normalizers of soluble groups developed by Professor P. Hall" (page 89 of [24]). He starts with the following theorem.

Theorem 5.15. [Theorem 2 in Carter [24]] *Let G be a solvable group of nilpotent length 2 in which L_1 is abelian. Then, the system normalizers of G complement L_1 and every complement to L_1 is a system normalizer.*

Carter's Proof. Since G/L_1 is nilpotent, it follows from Theorem 5.6 that $G = L_1\mathcal{N}_G(\Sigma)$ for a Sylow system Σ of G. Let $M = L_1 \cap \mathcal{N}_G(\Sigma)$. Given that L_1 is abelian, it follows that $M \trianglelefteq G$. Suppose that $M \neq \{1\}$ and let N be minimal normal in G such that $N \leq M$. Then, by Theorem 1.40, $N \leq Z_\infty(G)$. Since all of the chief factors in $Z_\infty(G)$ are central, this means that $N \leq Z(G)$. However, this means that $L_1 \cap Z(G) \neq \{1\}$, which is a contradiction (in Theorem 1 from [24] it is proven that if G be a solvable group of nilpotent length 2 with L_1 is abelian, then $L_1 \cap Z(G) = \{1\}$). Thus, $L_1 \cap \mathcal{N}_G(\Sigma) = \{1\}$ and $G = [L_1]\mathcal{N}_G(\Sigma)$.

Now let K be a complement of L_1 in G, and let p divide the order of G. Let S be a Hall p'-subgroup of K, which being nilpotent implies that $S \trianglelefteq K$. Let T be a Hall p'-subgroup of L_1, which being abelian implies that T is characteristic in G. Thus, ST is a Hall p'-subgroup of G with $K \leq \mathcal{N}_G(ST)$. Given that p was arbitrary, it then follows that $K \leq \mathcal{N}_G(\Sigma)$ for some Sylow system Σ of G. Given that $|K| = |\mathcal{N}_G(\Sigma)|$, $K = \mathcal{N}_G(\Sigma)$. □

Note that Carter's result also implies that all of the complements of L_1 are conjugate as the system normalizers of a group form a conjugacy class in G. Carter also gives a slightly improved version of G. Higman's theorem 5.14.

Theorem 5.16. [Theorem 4 in Carter [24]] *Let G be a solvable group of nilpotent length n in which L_{n-1} is abelian. Then, there are subgroups X of G satisfying $G = L_{n-1}X$ with $L_{n-1} \cap X = \{1\}$ and all such subgroups are conjugate. These subgroups are the relative system normalizers of L_{n-2} in G.*

Beweis. Since L_{n-2}/L_{n-1} is nilpotent, it follows from Theorem 5.6 that $G = L_{n-1}$ $\mathcal{N}_G(\Sigma_{L_{n-2}})$. Let $A = L_{n-2} \cap N_G(\Sigma_{L_{n-2}})$ be an absolute system normalizer of L_{n-2}. By Theorem 5.15, $L_{n-2} = [L_{n-1}]A$, which implies $L_{n-1} \cap A = \{1\}$. Thus,

$$L_{n-1} \cap \mathcal{N}_G(\Sigma_{L_{n-2}}) \leq L_{n-1} \cap A = \{1\}$$

and $G = [L_{n-1}]\mathcal{N}_G(\Sigma_{L_{n-2}})$.

Now let X be a complement to L_{n-1} in G. By Theorem 5.14, it follows that X and $\mathcal{N}_G(\Sigma_{L_{n-2}})$ are conjugate, which implies that X is also a relative system normalizer of L_{n-2} in G. □

Carter then applies Theorem 5.15 and Theorem 5.16 to A-groups. Recall that for A-groups, the derived series and lower nilpotent series coincide (Lemma 5.5) and that each subgroup in the lower nilpotent series is complemented (Corollary 5.9). He also showed (see Example 5.10) that not all complements of the subgroups in the lower nilpotent series need to be conjugate. However, Carter was able to prove the following result.

Theorem 5.17. [Theorem 5 in Carter [24]] *Let G be an A-group of length n. Then, all the complements of G^{n-1} are conjugate in G and all of the complements of G^i are conjugate "modulo G^{i+1}", i.e. if X and Y are complements of G^i, then $G^{i+1}X$ and $G^{i+1}Y$ are conjugate in G.*

He then goes on to say that "the class of groups for which L_i is complemented by relative system normalizers of L_{i-1} is, however, not confined to A-groups, but is a considerably wider class as shown in the following theorem" (page 93 in [24]).

Theorem 5.18. [Theorem 6 in Carter [24]] *If the lower nilpotent series of L_i coincides with the derived series, i.e. if the factors*

$$L_i/L_{i+1}, L_{i+1}/L_{i+2}, \ldots, L_{n-2}/L_{n-1}, L_{n-1}$$

are all abelian, then L_i is complemented in G by the relative system normalizers of L_{i-1}. In particular, if the lower nilpotent series coincides with the derived series, each L_i is complemented in G by the relative system normalizers of L_{i-1}.

In 1960, Newman [82] studied *JM*-groups. A group is **just metabelian** if it is metabelian (has a proper nontrivial abelian commutator subgroup), but every homomorphic image is abelian. Thus, a just metabelian group G has G' nontrivial and abelian with G' being the unique minimal normal subgroup of G. If $Z(G)$ is nontrivial, then $G' \leq Z(G)$ and G is nilpotent of class 2. Newman investigated this class of nilpotent groups in [83]. Newman was also interested in just metablelian groups with a trivial center. He defines a ***JM*-group** to be a just metabelian group with a trivial center. Newman's goal in [82] was to classify both finite and infinite *JM*-groups. He did prove some interesting results concerning complementation in these groups. Before proceeding, note that for a *JM*-group G, G' is the unique minimal normal subgroup of G and that $C_G(G') = G'$ (see Theorem 3.4 in [82]).

Theorem 5.19. [Theorem 3.3 in Newman [82]] *A JM-group splits over its commutator subgroup.*

Newman's Proof. Let G be a *JM*-group (recall that $G' \neq \{1\}$). Let $h \in G$ such that $h \notin C_G(G')$. Define a mapping $\alpha_h : G' \to G'$ by $\alpha_h(x) = [h, x]$ for $x \in G'$, which is an automorphism of G'.

Let $g \in G$ and note that $h^{-1}g^{-1}hg \in G'$. Given that the mapping α_h is an automorphism of G', there exists a unique element $b \in G'$ such that $h^{-1}g^{-1}hg = \alpha_h(b^{-1})$. This implies $h^{-1}g^{-1}hg = h^{-1}bhb^{-1}$ or that $h^{-1}g^{-1}hg = b^{-1}h^{-1}bh$ as G' is abelian. It then follows that $h^{-1}g^{-1}hgh^{-1}b^{-1}hb = 1$. This means $h^{-1}b^{-1}hh^{-1}g^{-1}hgb = 1$ or $h^{-1}b^{-1}g^{-1}hgb = 1$. As a result, $[h, gb] = 1$.

Thus, for each element $g \in G$, there is a unique element $b \in G'$ such that $gb \in C_G(h)$. Consequently, $G = G'C_G(h)$. Given that $h \notin C_G(G')$, $G' \cap C_G(h) = \{1\}$ and $G = [G']C_G(h)$. □

Note that having a trivial center is important as the following example indicates.

Example 5.20. Consider the group $G = \langle a, b, c \mid a^3 = b^3 = c^3 = 1, ab = ba, ac = ca, c^{-1}bc = ab \rangle$. This group G is just metabelian with the unique minimal normal subgroup $G' = \langle a \rangle$. However, G does not split over G', as it is also equal to $\Phi(G)$.

Newman also proved the following stronger result.

Theorem 5.21. [Theorem 3.5 in Newman [82]] *The complements of the commutator subgroup of a JM-group are the centralizers of the elements outside of the commutator subgroup. They are maximal subgroups and are all conjugate. There is a one-to-one correspondence between the elements of the commutator subgroup and its complements.*

Newman's Proof. Let G be a JM-group. The fact that the centralizers of elements outside of the commutator subgroup are the complements of G' follows from Theorem 5.19 and the fact that for any JM-group $C_G(G') = G'$.

Let K be a complement of G' in G. Then, $K \cong G/G'$ is abelian and $K \leq C_G(k)$ for any element $k \in K$. Then, by the modular identity (Lemma 1.6)

$$\begin{aligned} C_G(k) &= G \cap C_G(k) \\ &= KG' \cap C_G(k) \\ &= K(G' \cap C_G(k)). \end{aligned}$$

Given that $k \notin G'$, we know that $k \notin C_G(G')$. Thus, $C_G(k)$ is a complement to G' in G and $G' \cap C_G(k) = \{1\}$. This implies $C_G(k) = K$.

Suppose that K is not maximal in G and that $K < M < G$ for some subgroup M of G. Let $h \in K$, let $g \in G$, and let $f \in M - K$, then $g = cl$ for $c \in G'$ and $l \in K$. Thus,

$$\begin{aligned} [f, h]^g &= (f^{-1}h^{-1}fh)^g \\ &= g^{-1}f^{-1}h^{-1}fhg \\ &= l^{-1}c^{-1}f^{-1}h^{-1}fhcl \\ &= l^{-1}f^{-1}h^{-1}fhc^{-1}cl \\ &= l^{-1}f^{-1}h^{-1}fhl \\ &= l^{-1}f^{-1}lff^{-1}l^{-1}h^{-1}fhl \\ &= (f^{-1}l^{-1}fl)^{-1}[f, hl] \\ &= [f, l]^{-1}[f, hl], \end{aligned}$$

which implies that $[f, h]^g \in [f, K]$ and that $[f, K]$ is normal in G.

Recall that G' is a minimal normal subgroup of G. Thus, if $f \notin K$, then $G = G'K \leq [f, K]K$. This implies $G = [f, K]K$, which is a contradiction as $[f, K] \leq M$. All this implies that K is maximal in G.

Let K and L be two complements of G' in G. Thus, there are elements $k, l \in G$ such that $K = C_G(k)$ and $L = C_G(l)$. By the work done in the proof of Theorem 5.19, the mappings α_k, where $\alpha_k(y) = [k, y]$, and α_l, where $\alpha_l(y) = [l, y]$, for $y \in G'$, are automorphisms of G'. Given that $[l, k] \in G'$, there exists an element $x \in G'$ such that $\alpha_l(\alpha_k(x)) = [l, k]$. This implies

$$\begin{aligned} \alpha_l(k^{-1}x^{-1}kx) &= [l, k] \\ x^{-1}k^{-1}xkl^{-1}k^{-1}x^{-1}kxl &= [l, k] \\ [k, x, l] &= [l, k] \end{aligned}$$

or that $[l, k]^{-1}\ [k, x, l] = 1$. Now consider $[k^x, l]$. A simple calculation shows

$$\begin{aligned} [k^x, l] &= x^{-1}k^{-1}xl^{-1}x^{-1}kxl \\ &= x^{-1}k^{-1}xkk^{-1}l^{-1}kll^{-1}k^{-1}x^{-1}kxl \\ &= k^{-1}l^{-1}klx^{-1}k^{-1}xkl^{-1}k^{-1}x^{-1}kxl \\ &= [l, k]^{-1}[k, x, l] \\ &= 1. \end{aligned}$$

Consequently $k^x \in L$.

Since G' is abelian, it follows for and element $g \in G$ that (see II.6 (17) in [114])

$$[k^x, l, g^x][l, g^x, k^x][g^x k^x, l] = 1.$$

If $g \in K$, then $gk = kg$, with a simple calculation showing $[l, g^x, k^x] = 1$. Thus, $[k^x, l, g^x][g^x k^x, l] = 1$, which results in $[l, g^x] = 1$. This means $g^x \in L$ and that $K^x \leq L$. Given that K and L are maximal subgroups, it follows that $K^x = L$. Given that x is unique, there is a one-to-one correspondence between the elements of G' and the complements of G'. □

In 1964, the results by Newman were generalized by Yonaha [112] to metabelian groups.

Theorem 5.22. [Theorem 1 in Yonaha [112]] *Let G be a metabelian group whose commutator subgroup G' contains no nontrivial central elements ($G' \cap Z(G) = \{1\}$). Then*
(i) *G splits over G' and*
(ii) *the complements of G' in G form a conjugacy class.*

Given that the proof of Theorem 5.22 is lengthy and contains elements of the proofs of Theorem 5.19 and Theorem 5.21, it is omitted. Theorem 5.19 is a direct result from Theorem 5.22 as *JM*-groups have a trivial center.

This section ends with a result by Hawkes. A group G is **primitive** if it has a faithful primitive permutation representation. Equivalent to this condition is that G has a maximal subgroup M such that Core(M) = $\{1\}$. If G is solvable, it has a unique minimal normal subgroup N such that $G = [N]M$. For more information on primitive groups, see [87, 96]. A group G is said to be **multiprimitive** if for each normal subgroup N of G, G/N is primitive. Multiprimitive groups were introduced and studied by Hawkes [58], where he showed that an arbitrary finite solvable group can be embedded in a solvable multiprimitive group. He established the following result.

Theorem 5.23. [Theorem 1.1 in Hawkes [58]] *Each of the following is necessary and sufficient for a nontrivial solvable group G to be multiprimitive.*
(i) *G has a unique chief series and all of its chief factors are complemented.*
(ii) *The lower nilpotent series of G is a chief series.*
(iii) *The upper nilpotent series of G is a chief series.*
(iv) *G is extreme and G has prefrattini subgroups of order 1.*

Note that a group G is **extreme** if and only if $\mathcal{F}_i(G)/\Phi_i(G)$ (with $\mathcal{F}_i(G)$ and $\Phi_i(G)$ defined in the natural way) is a chief factor of G for each i, where $1 \leq i \leq r$ with r the nilpotent length of G. More information can be found in [25].

6 Normal subgroups with abelian sylow subgroups

In Chapter 5, a number of normal subgroup complementation results were presented for solvable groups with abelian Sylow subgroups (A-groups). This chapter presents conditions for when normal subgroups with abelian Sylow subgroups have complements. The chapter starts with a simple definition/lemma.

Lemma 6.1. *Let p divide the order of a group G. There exists a unique normal subgroup $G'(p)$ of G, the* ***p*-commutator subgroup of *G***, *such that*
(i) *$G/G'(p)$ is an abelian p-group, and*
(ii) *if G/A is an abelian p-group, then $G'(p) \leq A$.*

Beweis. Let $N = \bigcap_i \{N_i \mid N_i \trianglelefteq G$ and G/N_i is an abelian p-group$\}$. Given that G/N_i is abelian, $G' \leq N_i$ and G/N is abelian. Furthermore, since $G/(N_i \cap N_j)$ is isomorphic to a subgroup of $G/N_i \times G/N_j$, it follows that G/N is a p-group. Thus, $G'(p) = N$ is the unique normal subgroup of G such that $G/G'(p)$ is an abelian p-group. It also follows that if G/A is an abelian p-group that $G'(p) \leq A$. □

It follows directly that

$$\bigcap_{p \in |G|} G'(p) = G'$$

and that $O^p(G) \leq G'(p)$. The first main result in this chapter is due to Gaschütz.

Theorem 6.2. [Satz 7 in Gaschütz [41]] *For a group G, if the Sylow p-subgroups of $O^p(G)$ are abelian, then G splits over $O^p(G)$.*

The proof presented here is the one given by Rose [88]. The goal of Rose's paper was to present a proof of Theorem 6.2 that did not require extension theory.

Rose's Proof. Proceed by induction on the order of G. Let $K = O^p(G)$, and let P be a Sylow p-subgroup of K. Since P is abelian, $Z(P) = P$. Furthermore, if P^* is another Sylow p-subgroup of K that contains $Z(P)$, then $P = P^*$ and $Z(P) = Z(P^*)$. Also note that since $O^p(K)$ is characteristic in K, it will be normal in G, which implies that $O^p(K) = K$.

By the second theorem of Grün (see 8.3.5 in [7]),

$$K/K'(p) \cong \mathcal{N}_K(Z(P))/\mathcal{N}_K(Z(P))'(p) = \mathcal{N}_K(P)/\mathcal{N}_K(P)'(p).$$

Given that $O^p(K) = K$, it follows that $K'(p) = K$, which implies $\mathcal{N}_K(P) = \mathcal{N}_K(P)'(p)$. This means $O^p(\mathcal{N}_K(P)) = \mathcal{N}_K(P)$.

DOI 10.1515/9783110480214-006

Let $N = \mathcal{N}_G(P)$, which yields $N \cap K = \mathcal{N}_K(P)$. Given that N and K are normal in G, $N \cap K \lhd G$ and $N \cap K \trianglelefteq N$. Since $NK/K \leq G/K$, which is a p-group, and $|N/N \cap K| = |NK/K|$, it follows that $N/N \cap K$ is a p-group. Since $O^p(\mathcal{N}_K(P)) = \mathcal{N}_K(P)$ and $\mathcal{N}_K(P) = N \cap K$, a direct consequence is $O^p(N \cap K) = N \cap K$. Given that $N/N \cap K$ is a p-group, $O^p(N) \leq N \cap K$. However, since $O^p(N \cap K) = N \cap K$, it follows that $O^p(N) = N \cap K$. However, $N \cap K \leq K = O^p(G)$, which has abelian Sylow p-subgroups, implying $N \cap K$ has abelian Sylow p-subgroups.

If $N < G$, then by the induction hypothesis, N splits over $O^p(N) = N \cap K$ or $N = [N \cap K]H$, where $H \leq N$ and is a p-group. First note that $K \cap H = \{1\}$ since $N \cap K \cap H = \{1\}$. Second, by the Frattini argument (Lemma 1.7), note that $G = KN$. Let $g \in G$. Then, $g = kn$ for $k \in K$ and $n \in N$. Since $n = n_1h$, where $n_1 \in N \cap K$ and $h \in H$, it follows that $g = kn = kn_1h = n_2h$, with $n_2 = kn_1 \in K$. This means $G = KH$ or that $G = [K]H$.

Now consider the case where $N = \mathcal{N}_G(P) = G$. This means $P \lhd G$ and that $P \lhd K$. By Theorem 2.12, $K = [P]Q$ for Q a Hall p'-subgroup of K. Given that $O^p(K) = K$, the normal closure of Q in K is K or $Q^K = K$. By Lemma 2.15, $\mathcal{N}_K(Q) = Q$. Then, by Lemma 2.16, $G = K\mathcal{N}_G(Q)$. Let P_1 be a Sylow p-subgroup of $\mathcal{N}_G(Q)$. Since $Q \trianglelefteq \mathcal{N}_G(Q)$, it follows that $\mathcal{N}_G(Q) = QP_1$. The fact that $Q \leq K = O^p(G)$ and $G = K\mathcal{N}_G(Q)$ results in $G = KP_1$. By the modular identity (Lemma 1.6)

$$Q = \mathcal{N}_K(Q) = K \cap \mathcal{N}_G(Q) = K \cap QP_1 = Q(K \cap P_1).$$

Since $K \cap P_1$ is a p-subgroup of G contained in Q, which is a p'-subgroup of G, $K \cap P_1 = \{1\}$ and $G = [K]P_1$. □

Now come a series of results by Šemetkov. His motivation was to generalize and extend Theorem 6.2 by Gaschütz and Theorem 5.7 by Hall. To establish the first of these results, Šemetkov first needed a slightly different version of Theorem 4.9. Recall that a subgroup N of a group G has a π-complement H if $G = NH$ and $|N \cap H|$ is not divisible by any primes in π.

Theorem 6.3. [Theorem 4 in Šemetkov [99]] *A normal subgroup N of a group G has a π-complement in G if a Sylow p-subgroup of N is abelian and $O^p(N) = N$ for each $p \in \pi$.*

Šemetkov's Proof. Let P be a Sylow p-subgroup of G and consider the group PN. Given that $PN/N \cong P/N \cap P$, PN/N is a p-group, implying that $O^p(PN) \leq N$. However, given that $O^p(N) = N$, it follows that $O^p(PN) = N$. Thus, by Theorem 6.2, $PN = [N]H$, where H is a p-subgroup of PN.

Now $H \leq P_1$, a Sylow p-subgroup of PN. Thus, there exists an element $x \in PN$ such that $P_1^x = P$. By Corollary 1.13, $PN = [N]H^x$, where $H^x \leq P$. Applying the modular identity (Lemma 1.6), it follows that

$$P = P \cap PN = P \cap NH^x = (P \cap N)H^x,$$

where $P \cap N \cap H^x = \{1\}$. Thus, $P \cap N$ is abelian and complemented in P. By Theorem 4.9, N has a π-complement in G. □

Theorem 6.3 was used by Šemetkov to prove a conjecture he made in [98] which also appears as Question 3.59 in [76].

Theorem 6.4. [Šemetkov] *Let N be a non-solvable minimal normal subgroup of a finite group G, and suppose that N has cyclic Sylow p-subgroups for every prime dividing* $[G : N]$. *Then, N has a complement in G.*

Beweis. Let π be the set of primes dividing $[G : N]$. If each prime p that divides the order of N is also in π, then all of the Sylow p-subgroups of N are cyclic. In this case N is solvable (see page 160 of [67]), which is a contradiction.

If $(|N|, [G : N]) = 1$, then N has a complement in G by Theorem 2.12. Thus, not all primes in π divide the order of N and there exists primes that divide the order of N that are in π. If $p \in \pi$ and p does not divide the order of N, then trivially the Sylow p-subgroups of N are abelian with $O^p(N) = N$. If $p \in \pi$ and p divides the order of N, the Sylow p-subgroup of N is cyclic and it must be abelian. Furthermore, since N is minimal normal in G and $O^p(N)$ is characteristic in N, it must be that $O^p(N) = N$. Thus, the conditions of Theorem 6.3 are met and N has a π complement H. That is, $G = NH$ and $|N \cap H|$ is not divisible by any primes in π.

Furthermore, since $|NH/N| = |H/N \cap H|$, the order of $H/N \cap H$ is a π-number. By Lemma 2.17, G splits over N. □

At this point, Šemetkov's generalization of Theorem 6.2 can be presented.

Theorem 6.5. [Theorem 5 in Šemetkov [99]] *If a Sylow p-subgroup of* $O^\pi(G)$ *is abelian for each p in* π, *then* $O^\pi(G)$ *is complemented in G.*

Beweis. Clearly, $O^p(O^p(G)) = O^p(G)$ for each $p \in \pi$. By an argument similar to the one given in the proof of Theorem 6.3, it can be shown for each $p \in \pi$ and each Sylow p-subgroup P of G that $P \cap O^\pi(G)$ is abelian and complemented in P. Thus, by Theorem 6.3, $O^\pi(G)$ has a π-complement H in G. Then, by Lemma 2.17, $O^\pi(G)$ has a complement in G. □

To obtain his generalization of Hall's result (Theorem 5.7), Šemetkov's first generalizes the concepts of an A-group (see Definition 5.4) and the p-commutator subgroup of a group G (see Lemma 6.1).

Definition 6.6. A group in which a Sylow p-subgroup is abelian for any $p \in \pi$ is called an A_π**-group**. The π**-commutator** of a group G is the intersection of the normal subgroups N of G such that G/N is an abelian π-group.

Essentially, the π-commutator C of a group G is $C = \bigcap_{p\in\pi} G'(p)$. Before getting to the splitting result, yet another variation of Theorem 4.9 is needed.

Theorem 6.7. [Theorem 2.3 in Šemetkov [100]] *Suppose that a normal subgroup N of a group G is an A_π-group. Then, the π-commutator of N will have a π-complement in G.*

Šemetkov's Proof. Let C be the π-commutator of N, and let $S = \bigcap_i \{N_i \mid N_i \trianglelefteq N$ and N/N_i is a solvable π-group$\}$. Note that $S \leq C$ and that S is characteristic in N. Let $p \in \pi$, and let P be a Sylow p-subgroup of G. Then, SP is a subgroup of G, with SP/S being a p-group. This implies that $O^p(SP) = S$. Thus, by Theorem 6.2, $SP = [S]P_1$ for $P_1 \leq SP$. Now $P_1 \leq P^*$, a Sylow p-subgroup of SP. Thus, there exists an element $x \in SP$ such that $(P^*)^x = P$, which implies that $P_1^x \leq P$. By Corollary 1.13, $SP = [S]P_1^x$. By the modular identity (Lemma 1.6),

$$P = SP \cap P = SP_1^x \cap P = (S \cap P)P_1^x.$$

Thus, $P = [S \cap P]P_1$, and by Theorem 4.9, S has a π-complement H in G.

Let $N_1 = N \cap H$, which is normal in H. Given that $N_1 \leq N$, N_1 is an A_π-group. Let C_1 be the π-commutator of N_1, and consider $N_1/N_1 \cap C$. Given that $N_1/N_1 \cap C \cong N_1C/C$, which is a subgroup of the abelian π-group N/C, it follows that $C_1 \leq N_1 \cap C$. Given that

$$N_1S = (N \cap H)S = N \cap HS = N \cap G = N$$

and that $S \cap N_1 = S \cap N \cap H = S \cap H$, S is a π-complement to N_1 in N. Since $N_1 \cap S$ is a π'-group and N_1/C_1 is an abelian π-group, $N_1 \cap C_1S = C_1(N_1 \cap S) = C_1$. In addition, with

$$N/C_1S = N_1C_1S/C_1S \cong N_1/N_1 \cap C_1S = N_1/C_1,$$

it follows that $C \leq C_1S$. Furthermore, $N_1 \cap C \leq N_1 \cap C_1S = C_1$, implying that $C_1 = N_1 \cap C = N \cap H \cap C = H \cap C$.

There are now three cases to examine.

Case 1: $|S|$ *is divisible by primes in* π. Given that $G = SH$, with $S \cap H$ being a π'-group, it follows that S is not contained in H and that $H \neq G$. Proceeding by induction on the order of G, C_1 has a π-complement L in H. Thus, $H = C_1L = (H \cap C)L = CL \cap H$, which implies in $H \leq CL$. As a result, since $S \leq C$, $G = SH = S(CL) = CL$, where $C \cap L = C \cap L \cap L \leq C \cap H \cap L \leq C_1 \cap L$. Thus, L is a π-complement to C in G.

Case 2: $|S| \neq 1$ *and is not divisible by primes in* π. It immediately follows that the order of N/S is the largest π-divisor of the order of N. Proceed by induction on the order of G. Now N/S is an A_π-group with the π-commutator subgroup C/S. Thus, C/S has a

π-complement L/S in G/S. This implies $G = CL$, with $C \cap L = S$. Since S is a π'-group, C has a π-complement in G.

Case 3: $|S| = 1$. In this case, N is a solvable π-group with abelian Sylow subgroups. Also note that in this case $C = N'$, the derived subgroup of N. By Lemma 1.33, $G = N\mathcal{N}_G(\Sigma_N)$, where Σ_N is a Sylow system for N. Let $T = \mathcal{N}_G(\Sigma_N) \cap N$, which is system normalizer for N. By Theorem 5.7, $N = [C]T$. This implies

$$G = N\mathcal{N}_G(\Sigma_N) = CT\mathcal{N}_G(\Sigma_N) = C\mathcal{N}_G(\Sigma_N),$$

where

$$C \cap \mathcal{N}_G(\Sigma_N) = C \cap N \cap \mathcal{N}_G(\Sigma_N) = C \cap T = \{1\}.$$

Thus, $\mathcal{N}_G(\Sigma_N)$ is a complement to C in G. □

Now Šemetkov's main result can be proven.

Theorem 6.8. [Theorem 2.4 in Šemetkov [100]] *Suppose π contains all prime divisors of the index $[G : N]$ of a normal subgroup N of G. If N is an A_π-group, then its π-commutator will have a complement in G*

Šemetkov's Proof. By Theorem 6.7, it follows that the π-commutator C of N has a π-complement H in G. Given that $H \cap C$ is normal in H, with $H/H \cap C \cong HC/C = G/C$, where G/C is a π-group, the result follows by Lemma 2.17. □

The next result is a direct application of Theorem 6.8.

Theorem 6.9. [Theorem 2.5 in Šemetkov [100]] *A normal subgroup N of the group G has a complement in G if N is an A_π-group that coincides with its π-commutator and G/N is a π-group.*

Now let π be the set of all primes. In this case, the π-commutator of a group coincides with its commutator subgroup. Let N be a normal subgroup of a group G with abelian Sylow subgroups. Then, by Theorem 6.8, N' will have a complement in G. This leads to the following result, which is Šemetkov's final generalization of Theorem 5.7 by Hall.

Theorem 6.10. [Theorem 2.6 in Šemetkov [100]] *If N is a normal subgroup of a group G and all the Sylow subgroups of N are abelian, then there exists in the group G a complement S of the commutator subgroup of N.*

7 The formation generation

The use of formations to obtain normal subgroup complementation results began with Shult [101], who used formations to generalize the results of Schenkman and G. Higman. Recall that Schenkman proved (Theorem 5.13) that if the hypercommutator $D_\infty(G) = L_1(G)$ of a group G is abelian, then it is complemented and all its complements are conjugate. G. Higman extended this result (Theorem 5.14) showing that if $L_r(G)$ is abelian for any integer r, then it is complemented and all of its complements are conjugate.

Consider the formation $\mathfrak{N}$ of nilpotent groups. For a group G, $L_1(G) = G_{\mathfrak{N}}$, where $G_{\mathfrak{N}}$ is the $\mathfrak{N}$-residual of G. The $\mathfrak{N}$-projectors of G are the Carter subgroups of G. Shult uses this fact to generalize Schenkman's and G. Higman's results as follows:

Theorem 7.1. [Main Theorem in Shult [101]] *Let $\mathfrak{F}$ be a saturated formation and suppose for a solvable group G that $G_{\mathfrak{F}}$ is abelian. Then*
(i) *The $\mathfrak{F}$-projectors of G complement $G_{\mathfrak{F}}$, and*
(ii) *All complements of $G_{\mathfrak{F}}$ in G are conjugate and hence $\mathfrak{F}$-projectors of G.*

Shult's Proof. If $G_{\mathfrak{F}} = \{1\}$, then G is the unique $\mathfrak{F}$-projector of G and a complement of $G_{\mathfrak{F}}$. Assume that $G_{\mathfrak{F}} \neq \{1\}$ and proceed by induction on the order of G. Let F be an $\mathfrak{F}$-projector of G. Given that $F \le G$ with $G_{\mathfrak{F}} \lhd G$ such that $G/G_{\mathfrak{F}} \in \mathfrak{F}$, it follows by the definition of an $\mathfrak{F}$-projector that $G = G_{\mathfrak{F}}F$.

To show that (i) holds, it must be shown that $G_{\mathfrak{F}} \cap F = \{1\}$. Suppose that $G_{\mathfrak{F}} \cap F \neq \{1\}$. Since $G_{\mathfrak{F}}$ is abelian, $G_{\mathfrak{F}} \cap F \lhd G$. Let N be a minimal normal subgroup of G. By Lemma 1.93, $(G/N)_{\mathfrak{F}} = G_{\mathfrak{F}}N/N$ and is abelian. Furthermore, Lemma 1.86 implies that FN/N is an $\mathfrak{F}$-projector of G/N. Thus, $G/N = (G/N)_{\mathfrak{F}}FN/N$ and, by induction, $(G/N)_{\mathfrak{F}} \cap FN/N = \{1_{G/N}\}$. This implies $G_{\mathfrak{F}}N \cap FN = N$ or that $G_{\mathfrak{F}} \cap F \le N$. Given that the choice of N was arbitrary, $G_{\mathfrak{F}} \cap F = \{1\}$ unless G has a unique minimal normal subgroup.

Proceed assuming G has a unique minimal normal subgroup N^* with $G_{\mathfrak{F}} \cap F = N^*$. If $G_{\mathfrak{F}} = N^*$, then $G = G_{\mathfrak{F}}F = N^*F = F$, which implies (recall $F \in \mathfrak{F}$) that $G_{\mathfrak{F}} = \{1\}$. This contradiction implies $N^* < G_{\mathfrak{F}}$.

Suppose that F is not maximal in G and that there exists a subgroup H of G such that $F < H < G$. By Lemma 1.93, $H_{\mathfrak{F}} \le G_{\mathfrak{F}}$ and is abelian. By Lemma 1.86, F is an $\mathfrak{F}$-projector of H and $H = H_{\mathfrak{F}}F$. By induction $H_{\mathfrak{F}} \cap F = \{1\}$. Given that $G = G_{\mathfrak{F}}F$ and $G_{\mathfrak{F}}$ is abelian, $H_{\mathfrak{F}} \cap F \lhd G$. However, given that $H_{\mathfrak{F}}$ is normal in G and that $N^* \le F$, it follows that $N^* \le H_{\mathfrak{F}} \cap F$, a contradiction. Thus, F is maximal in G.

Suppose there exists a normal subgroup N_1 of G such that $N^* \lhd N_1 \lhd G_{\mathfrak{F}}$. Since $G_{\mathfrak{F}} \cap F = N^*$, $N_1 \nleq F$ and $G = N_1F$. Thus, $G/N_1 = N_1F/N \cong F/N_1 \cap F$, where $F/N_1 \cap F \in \mathfrak{F}$. This contradiction implies

$$\{1\} \lhd N^* \lhd G_{\mathfrak{F}}$$

DOI 10.1515/9783110480214-007

is part of a chief series for G. Given that $G_{\mathfrak{F}}$ is abelian and N^* is the unique minimal normal subgroup of G, then $G_{\mathfrak{F}}$ must be an abelian p-group for some prime p.

Let Q be a p'-subgroup of G such that $QG_{\mathfrak{F}} \trianglelefteq G$. Since $G_{\mathfrak{F}}$ is abelian, $G_{\mathfrak{F}} = C_{G_{\mathfrak{F}}}(Q) \times [Q, G_{\mathfrak{F}}]$ (see 5.2.3 in [46]) with both $C_{G_{\mathfrak{F}}}(Q)$ and $[Q, G_{\mathfrak{F}}]$ normal in G. If $[Q, G_{\mathfrak{F}}] < G_{\mathfrak{F}}$, then, given that N^* is the unique minimal normal subgroup of G, it follows that $C_{G_{\mathfrak{F}}}(Q) = G_{\mathfrak{F}}$. This implies $QG_{\mathfrak{F}} = Q \times G_{\mathfrak{F}}$. Given that Q is characteristic in $QG_{\mathfrak{F}}$, Q is normal in G. This contradictions the uniqueness of N^*. Thus, either $Q = \{1\}$ or $[Q, G_{\mathfrak{F}}] = G_{\mathfrak{F}}$.

Let q be a prime. Let $L_q = O_{q'q}(G/N^*) = O_q((G/N^*)/O_{q'}(G/N^*))$, and let $F_q = O_{q'q}(F) = O_q(F/O_{q'}(F))$. It will be shown that $F_q \leq L_q$. If $q \neq p$, then since $G_{\mathfrak{F}}$ is an abelian p-group, $G_{\mathfrak{F}} \leq O_{q'}(G)$. Given that $O_{q'}(F)$ is characteristic in F, $G_{\mathfrak{F}}$ is normal in G, and $G = G_{\mathfrak{F}}F$, it follows that $F_qG_{\mathfrak{F}}$ is q-nilpotent and normal in G. Since N^* is a p-group, this implies

$$F_qG_{\mathfrak{F}} \leq O_{q'q}(G) \leq L_q$$

or that $F_q \leq L_q$.

Now suppose that $q = p$. Let $Q^* = O_{p'}(F)$. Given that $Q^*G_{\mathfrak{F}}$ is normal in G, it follows by an earlier observation that $Q^* = \{1\}$ or $[Q^*, G_{\mathfrak{F}}] = G_{\mathfrak{F}}$. Assume that $[Q^*, G_{\mathfrak{F}}] = G_{\mathfrak{F}}$. Since Q^* is normal in F and $N^* \leq O_p(F)$ is also normal in F, it follows that $[Q^*, N^*] = \{1\}$. This contradictions the assumption $[Q^*, G_{\mathfrak{F}}] = G_{\mathfrak{F}}$. Thus, $Q^* = \{1\}$ and

$$F_p = O_{p'p}(F) = O_p(F/O_{p'}(F)) = O_p(F).$$

Since $F_pG_{\mathfrak{F}}$ is normal in G and N^* is a p-group, $F_pG_{\mathfrak{F}} \leq O_p(G) \leq L_p$ or $F_p \leq L_p$.

By Theorem 1.91, $\mathfrak{F}$ is locally defined by $\{\mathfrak{F}_p\}$, where each $\mathfrak{F}_p \subseteq \mathfrak{F}$. Let q divide the order of F. Since $F \in \mathfrak{F}$, it follows that $F/F_q \in \mathfrak{F}_q$ (recall that $F_q = O_{q'q}(F) = \cap\{C_F(H/K) \mid \text{where } H/K \text{ is a } q\text{-chief factor in } F\}$). Given that $F_q \leq L_q$,

$$(F/F_q)/(F \cap L_q/F_q) \cong F/F \cap L_q \in \mathfrak{F}_q.$$

Since $G_{\mathfrak{F}} \leq L_q$, the calculation

$$G/L_q = L_qF/L_q \cong F/F \cap L_q$$

shows that $G/L_q \in \mathfrak{F}_q$. Since F is maximal in G with $G = G_{\mathfrak{F}}F$, this also shows that $G/L_q \in \mathfrak{F}_q$ for each q dividing $[G : N^*]$. Thus, $G/N^* \in \mathfrak{F}$, which is a contradiction as $N^* < G_{\mathfrak{F}}$. Thus, $G_{\mathfrak{F}} \cap F = \{1\}$ and $G = [G_{\mathfrak{F}}]F$.

For (ii), let K be a complement to $G_{\mathfrak{F}}$ in G and assume that $G_{\mathfrak{F}} \neq \{1\}$. Proceed by induction on the order of G. Let N be a minimal normal subgroup of G

such that $N \leq G_{\mathfrak{F}}$. Then, by Lemma 1.11 and Lemma 1.93, $G/N = [G_{\mathfrak{F}}/N]KN/N = [(G/N)_{\mathfrak{F}}]KN/N$. By induction KN/N is an $\mathfrak{F}$-projector of G/N. Let F be an $\mathfrak{F}$-projector of G. Then, by Lemma 1.86, FN/N is an $\mathfrak{F}$-projector of G/N. Thus, by Theorem 1.88, KN/N and FN/N are conjugate in G/N. By Lemma 1.14, KN and FN are conjugate in G and there exists an element $g \in G$ such that $KN = (FN)^g$ or that $KN = F^gN$.

By Lemma 1.86, F^g is an $\mathfrak{F}$-projector of G. Now suppose that $N < G_{\mathfrak{F}}$. Then, $KN = F^gN < G$. Furthermore, since (using Corollary 1.13) $G = [G_{\mathfrak{F}}]F^g = [G_{\mathfrak{F}}]K$, $F^g \cap N \leq F^g \cap G_{\mathfrak{F}} = \{1\}$ and $K \cap N \leq K \cap G_{\mathfrak{F}} = \{1\}$ with $F^g < F^gN$ and $K < KN$. Therefore $(KN)_{\mathfrak{F}} = (F^gN)_{\mathfrak{F}} = N$ and $KN = [(KN)_{\mathfrak{F}}]F^g = [(KN)_{\mathfrak{F}}]K$ with F^g and K being $\mathfrak{F}$-projectors of KN. By induction F^g and K are conjugate in KN and are thus conjugate in G. By Lemma 1.86, K is an $\mathfrak{F}$-projector for G.

Finally consider the case that $G_{\mathfrak{F}} = N$. Thus, $G/G_{\mathfrak{F}} \cong K$ and $K \in \mathfrak{F}$ and is maximal in G. This implies that K is an $\mathfrak{F}$-projector of G. □

Now credit for Theorem 7.1 also needs to be given to Carter and Hawkes. As mention in Section 1.8, they introduced the concept of an $\mathfrak{F}$-normalizer. They also showed that given a solvable group G and a saturated formation $\mathfrak{F}$, that if $G/N \in \mathfrak{F}$ for N a normal nilpotent subgroup of G, then the $\mathfrak{F}$-normalizers and the $\mathfrak{F}$-projectors of G coincide. Given that both they and Schult are assuming that the $\mathfrak{F}$-residual is abelian (and thus nilpotent), Carter and Hawkes obtain the following result which is identical to Shult's result (Theorem 7.1).

Theorem 7.2. [Theorem 5.15 in Carter and Hawkes [26]] *Let $\mathfrak{F}$ be a local formation, and let G be a solvable group where the $\mathfrak{F}$-residual $G_{\mathfrak{F}}$ is abelian. Then, $G_{\mathfrak{F}}$ is complemented and any two complements are conjugate. The complements are the $\mathfrak{F}$-normalizers of G.*

In general, when $\mathfrak{F}$ is a saturated formation and G a solvable group, the $\mathfrak{F}$-residual $G_{\mathfrak{F}}$ need not have a complement in G as demonstrated by the following example due to Seitz and Wright [97].

Example 7.3. Let group G be the central product of $SL_2(3)$ with C_4, where the subgroups of order 2 from $SL_2(3)$ and C_4 are amalgamated. The $G_{\mathfrak{N}}$-residual, for $\mathfrak{N}$ the formation of nilpotent groups, is isomorphic to the quaternion group Q_8 and has no complement in G.

Cigić [30] showed with the following example that there are groups which have $G_{\mathfrak{F}}$ nilpotent, but where the $\mathfrak{F}$-projectors do not complement $G_{\mathfrak{F}}$. This example also shows that when $\mathfrak{F}$ is a saturated formation and G a solvable group, if the $\mathfrak{F}$-residual $G_{\mathfrak{F}}$ has a complement, it need not be an $\mathfrak{F}$-projector.

Example 7.4. Let $G = \langle a, b, x \mid a^4 = b^4 = x^3 = 1, a^2 = b^2, ab = ba^3, ax = xb, bx = xab\rangle$. Essentially, $G = [Q_8]C_3$, where Q_8 is the quaternion group of eight elements. For the formation $\mathfrak{N}$ of nilpotent groups, $G_{\mathfrak{N}} = \langle a, b\rangle$, which has a complement $\langle x\rangle$ in G. However, $F = \langle a^2, x\rangle$ is the $\mathfrak{N}$-projector of G and F does not complement $G_{\mathfrak{N}}$.

The next natural step is to weaken the abelian condition on the $\mathfrak{F}$-residual $G_{\mathfrak{F}}$ and still obtain a complement to $G_{\mathfrak{F}}$ in G. In 1970, Seitz and Wright [97] were able weakened the condition, as stated by Shult in Theorem 7.1 and Carter and Hawkes in Theorem 7.2, for when the $\mathfrak{F}$-residual $G_{\mathfrak{F}}$ has a complement in G. Their generalization of these two theorems has the same flavor as the generalization by Šemetkov in Theorem 4.10 of Gaschütz's Theorem 4.4. Note that a 2-group G is **quaternion free** if there do not exist normal subgroups H and K of G such that H/K is isomorphic to the quaternion group of order 8. In addition, recall that $\mathfrak{S}_\pi$ is the formation of π-groups and that a p-group is modular (has a modular subgroup lattice) if and only if any two of it subgroups permute (see Corollary 1.74).

Theorem 7.5. [Theorem 2.2 in Seitz and Wright [97]] *Let $\mathfrak{F}$ be a saturated formation, and let G be a solvable group. Let τ be the set of primes dividing $[G : G_{\mathfrak{F}}]$. Suppose that for each $p \in \tau$ that the Sylow p-subgroups of $G_{\mathfrak{F}}$ are quaternion free if $p = 2$ and are modular if p is odd. Let D be an $\mathfrak{F}$-normalizer of G and D_τ a Hall τ-subgroup of D. Then, $G = G_{\mathfrak{F}}D_\tau$ and $G_{\mathfrak{F}} \cap D_\tau = \{1\}$. Moreover, D_τ is an $(\mathfrak{F} \cap \mathfrak{S}_\tau)$-normalizer of G.*

To prove this theorem, two results are needed. They stem from the classic result that if P is an abelian Sylow p-subgroup of a group G, then $P \cap Z(G) \cap G' = \{1\}$ (Satz 3.2 in [65]). The machinery required to complete their proofs is extensive and extends well beyond the scope of this book. Consequently, they are omitted.

Theorem 7.6. [Satz 3.2 in Huppert [64]] *If G is a group, if p is an odd prime, and if P is a modular Sylow p-subgroup of G, then $P \cap Z(G) \cap G_{\mathfrak{N}} = \{1\}$.*

Theorem 7.7. [Satz 2.8 in Dornhoff [35]] *If G is a solvable group with quaternion-free Sylow p-subgroups P, then $P \cap Z(G) \cap G_{\mathfrak{N}} = \{1\}$.*

Seitz and Wright's Proof. To obtain the first part of this result, all that needs to be shown is that all of the chief factors of G contained in $G_{\mathfrak{F}}$ are eccentric. If this is the case, then by Theorem 1.97 it follows that D avoids each chief factor of G in $G_{\mathfrak{F}}$. It then follows by Lemma 5.1 that $G_{\mathfrak{F}} \cap D = \{1\}$. Now by Corollary 1.100, $D \leq F$, where F is an $\mathfrak{F}$-projector for G. By definition, $G = G_{\mathfrak{F}}F$. Given that D_τ is a Hall τ-subgroup of G and $[G : G_{\mathfrak{F}}]$ is a τ-number, if follows that $G = G_{\mathfrak{F}}D_\tau$ with $G_{\mathfrak{F}} \cap D_\tau \leq G_{\mathfrak{F}} \cap D = \{1\}$.

To show that all chief factors of G in $G_{\mathfrak{F}}$ are eccentric, assume this is not the case and that G is a minimal counterexample. Let N be a minimal normal subgroup of G, where N is an abelian p-group. By Lemma 1.93, $(G/N)_{\mathfrak{F}} = G_{\mathfrak{F}}N/N$. This implies that all chief factors of G between N and $NG_{\mathfrak{F}}$ are $\mathfrak{F}$-eccentric. If $N \nleq G_{\mathfrak{F}}$, then $G_{\mathfrak{F}}N/G_{\mathfrak{F}} \cong N/N \cap G_{\mathfrak{F}} = N$ is $\mathfrak{F}$-eccentric. This means $N \leq G_{\mathfrak{F}}$ and that N is $\mathfrak{F}$-central. Now $Aut_G(N) \in \mathfrak{F}_p$, which implies $G/C_G(N) \in \mathfrak{F}_p$ or that $G/C_G(N) \in \mathfrak{F}$ (recall that by

Theorem 1.91 that $\mathfrak{F}$ is locally defined by $\{\mathfrak{F}_p\}$, where $\mathfrak{F}_p \subseteq \mathfrak{F}$ for each p). Thus, $G_{\mathfrak{F}} \leq C_G(N)$ or $N \leq Z(G_{\mathfrak{F}})$.

Suppose that $(G_{\mathfrak{F}})_{\mathfrak{N}} > \{1\}$. If $N \nleq (G_{\mathfrak{F}})_{\mathfrak{N}}$, then $N(G_{\mathfrak{F}})_{\mathfrak{N}}/N \cong N/(G_{\mathfrak{F}})_{\mathfrak{N}} \cap N = N$ and N is an eccentric p-chief factor. This again means $N \leq (G_{\mathfrak{F}})_{\mathfrak{N}}$. Consequently, $N \leq P \cap Z(G_{\mathfrak{F}}) \cap (G_{\mathfrak{F}})_{\mathfrak{N}}$, where P is a Sylow p-subgroup of $G_{\mathfrak{F}}$. By Theorem 7.6 and Theorem 7.7, this is a contradiction. Thus, $(G_{\mathfrak{F}})_{\mathfrak{N}} = \{1\}$ and $G_{\mathfrak{F}}$ is nilpotent. If $G_{\mathfrak{F}} \neq P$, then an argument similar to one presented twice earlier results in N being $\mathfrak{F}$-eccentric. This means $G_{\mathfrak{F}} = P$.

Let $Q/P = O_{p'p}(G/P)$. By Theorem 3.2 on page 216 in [106], $O_{p'p}(G/P) = \cap C_{G/N}((H/P)/(K/P))$ over all p-chief factors $(H/P)/(K/P)$ of G/P. This implies $(G/P)/(Q/P) \in \mathfrak{F}_p$ or that $G/Q \in \mathfrak{F}_p$. Since $G/C_G(N) \in \mathfrak{F}_p$, it follows that $G_{\mathfrak{F}_p} \leq Q \cap C_G(N)$.

Now Q/P has a normal Hall p'-subgroup, which implies $Q \cap C_G(N)/P$ has a normal Hall p'-subgroup. If $Q \cap C_G(N)$ is a p-group, then $G_{\mathfrak{F}_p}$ is also a p-group. This implies that for any q-chief factor H/K of G contained in $G_{\mathfrak{F}}$ that $G/C_G(H/K) \in \mathfrak{F}_q$ or that H/K is $\mathfrak{F}$-central. This contradiction implies that $Q \cap C_G(N)$ is not a p-group. Since $Q \cap C_G(N)/P$ has a normal Hall p'-subgroup, there exists a p'-chief factor H/K in $Q \cap C_G(N)/P$. This implies $N \leq Z(H) \cap P$. Applying Theorems 7.6 and 7.7 results in $H_{\mathfrak{N}} = \{1\}$. Thus, H contains a minimal normal subgroup for G that is a p'-group. This contradiction implies that all chief factors of G below $G_{\mathfrak{F}}$ are $\mathfrak{F}$-eccentric.

Now let A be an $(\mathfrak{F} \cap \mathfrak{S}_\tau)$-normalizer of G. Then, by the same method used earlier in the proof, A is a complement to $G_{\mathfrak{F}}$ in G. By Lemma 1.90, $\mathfrak{F} \cap \mathfrak{S}_\tau$ is locally defined by $\{\mathfrak{F}_p \cap \mathfrak{S}_\tau\}$.

Let Σ be a Sylow system for G. By definition, $A = \bigcap_{p\in\tau} \mathcal{N}_G(G_{\mathfrak{F}_p} \cap H_{p'})$, where $H_{p'}$ is the Hall p'-subgroup in Σ. Note that $D = \bigcap_p \mathcal{N}_G(G_{\mathfrak{F}_p} \cap H_{p'})$ for each prime p. Thus, $A \leq D \cap H_\tau$, where H_τ is a Hall τ-subgroup of G. As a result, A is contained in an $(\mathfrak{F} \cap \mathfrak{S}_\tau)$-normalizer of G. But since all normalizers are conjugate and A has the same order as D_τ, A is an $(\mathfrak{F} \cap \mathfrak{S}_\tau)$-normalizer of G. □

For a group G and the formation $\mathfrak{S}_\pi$, the $G_{\mathfrak{S}_\pi}$-residual is $O^\pi(G)$. Thus, Theorem 7.5 can be used to obtain the following generalization of Theorem 6.2 by Gaschütz and Rose and Theorem 6.5 by Šemetkov.

Corollary 7.8. [Corollary 2.3 in Seitz and Wright [97]] *Let G be a solvable group and π a set of primes. Suppose that for each $p \in \pi$ that the Sylow p-subgroups of $O^\pi(G)$ are modular for p odd and quaternion free for p even. Then, $O^\pi(G)$ is complemented by an $\mathfrak{S}_\pi$-normalizer of G.*

The next corollary of Theorem 7.5 is a generalization of Theorem 7.1 by Shult.

Corollary 7.9. [Corollary 2.4 in Seitz and Wright [97]] *Let $\mathfrak{F}$ be a saturated formation and let G be a solvable group. Suppose that $G_{\mathfrak{F}}$ has abelian Sylow subgroups for* *the primes dividing* $[G : G_{\mathfrak{F}}]$. *Then,* $G_{\mathfrak{F}}$ *is complemented by a Hall subgroup of an* $\mathfrak{F}$*-normalizer of G.*

However, as the following example from [97] shows, all complements in this case do not need to be conjugate.

Example 7.10. Let H be the primitive solvable group of degree 8 and of order $168 = 2^3 \cdot 3 \cdot 7$. Then, by 10.5.21 in [96], H has a unique minimal normal subgroup of order 2^3. Let $G = H \times \langle a \rangle$, where $|a| = 2$. Thus, G has abelian Sylow subgroups. Let D_0 be a system normalizer of H. Then, D_0 has order 3 and centralizes an involution z of H. The subgroups $D_0 \times \langle a \rangle$ and $D_0 \times \langle az \rangle$ both complement $G_{\mathfrak{N}}$ but are not conjugate.

Example 7.10 motivates the following result, which is a variation of Shult's result (Theorem 7.1).

Theorem 7.11. [Theorem 3.1 in Seitz and Wright [97]] *Let* $\mathfrak{F}$ *be a saturated formation. Suppose that G is a solvable group in which* $G_{\mathfrak{F}}$ *is complemented by a* $\mathfrak{F}$*-projector subgroup. Then, all complements of* $G_{\mathfrak{F}}$ *are conjugate.*

Seitz and Wright's Proof. If $G_{\mathfrak{F}} = \{1\}$, the result is trivial. Assume that $G_{\mathfrak{F}} \neq \{1\}$ and let G be a counter example of minimal order. Let F be an $\mathfrak{F}$-projector of G such that $G = [G_{\mathfrak{F}}]F$, and let $H \leq G$ such that $G = [G_{\mathfrak{F}}]H$. Let N be a minimal normal subgroup of G such that $N \leq G_{\mathfrak{F}}$.

By Lemma 1.11, $G/N = [G_{\mathfrak{F}}/N]HN/N = [G_{\mathfrak{F}}/N]FN/N$, which by Lemma 1.93 implies that $G/N = [(G/N)_{\mathfrak{F}}]HN/N = [(G/N)_{\mathfrak{F}}]FN/N$. Thus, HN/N and FN/N are conjugate in G/N. However, since any conjugate of a complement of a normal subgroup is also a complement of that normal subgroup (Corollary 1.13) and all $\mathfrak{F}$-projectors are conjugate (Theorem 1.88), it can be assumed that $NH = NF = T$. By Lemma 1.86, F is an $\mathfrak{F}$-projector of T. Furthermore, since $N \cap F = \{1\}$, it follows that $T/N = NF/N \cong F/N \cap F = F$ or that $T/N \in \mathfrak{F}$. This means $T_{\mathfrak{F}} = N$.

If $T < G$, then H and F are conjugate in T, implying that they are conjugate in G. Thus, any complement of $G_{\mathfrak{F}}$ is an $\mathfrak{F}$-projector and all complements are conjugate. If $T = G$, the contradiction implies the results follows. □

This corollary follows directly.

Corollary 7.12. [Corollary 3.2 in Seitz and Wright [97]] *Let* $\mathfrak{F}$ *be a saturated formation. Let G be a solvable group and* π *the set of primes dividing* $[G : G_{\mathfrak{F}}]$. *Suppose that for each prime p in* π, $G_{\mathfrak{F}}$ *has a normal Hall* p'*-subgroup and that the Sylow p-subgroups of* $G_{\mathfrak{F}}$ *are modular for p odd and quaternion free for p even. Then,* $G_{\mathfrak{F}}$ *is complemented by an* $(\mathfrak{F} \cap \mathfrak{S}_\pi)$*-normalizer of G and all complements of* $G_{\mathfrak{F}}$ *are conjugate.*

The next result by Seitz and Wright imposes conditions on the action of an $\mathfrak{F}$-normalizer on $G_{\mathfrak{F}}$ to obtain a complementation result. It is yet another variation of Shult's result (Theorem 7.1). Its proof is lengthy and is omitted.

Theorem 7.13. [Theorem 3.3 in Seitz and Wright [97]] *Let $\mathfrak{F}$ be a saturated formation and let D be an $\mathfrak{F}$-normalizer of the solvable group G. Suppose that for each prime p the Sylow p-subgroups act fixed-point-freely by conjugation on each p'-chief factor of G lying below $G_{\mathfrak{F}}$. Then, $G_{\mathfrak{F}}$ is complemented by D. Moreover, if the Sylow 2-subgroups of D are quaternion-free, then all complements of $G_{\mathfrak{F}}$ are conjugate.*

Theorem 7.13 can be applied to obtain the following result.

Theorem 7.14. [Theorem 4.2 in Seitz and Wright [97]] *Let $\mathfrak{F}$ be a saturated formation and let G be a solvable group. Let D be an $\mathfrak{F}$-normalizer of G, and suppose that for each prime p the Sylow p-subgroups of D act fixed-point-freely by conjugation on every p'-chief factor of G. Then, D is a p-group for some prime p, D is a system normalizer of G complementing $G_{\mathfrak{N}}$, and $G_{\mathfrak{N}}$ has a normal Sylow p-subgroup.*

To extend the results already established in this chapter to arbitrary normal subgroups, Wright [111] introduces two new conditions that are generalizations of properties satisfied by $G_{\mathfrak{F}}$. With these conditions, Wright is able to extend the results from [97].

Definition 7.15. Let G be a solvable group and let N be a normal subgroup of G.
(i) ${}^pN$ is the subgroup of G defined by ${}^pN/N = O_{p'}(G/N)$.
(ii) $N r_p\, G$ if and only if ${}^pN$ centralizes no p-chief factor of G of the form N/K.
(iii) $N r\, G$ if and only if $N r_p\, G$ for each prime p dividing $[G : N]$.
(iv) $N e\, G$ if and only if ${}^pN$ centralizes no p-chief factor of G below N for each prime p dividing $[G : N]$.
(v) If Σ is a Sylow system of G and π a set of primes, then

$$D_{N,\Sigma,\pi} = H_\pi \cap \bigcap_{p\in\pi} \mathcal{N}_G({}^pN \cap H_{p'}),$$

where H_π is the Hall π-subgroup of Σ, $H_{p'}$ is the Hall p'-subgroup of Σ, and $D_{N,\Sigma} = D_{N,\Sigma,\pi}$ when π is the set of primes dividing $[G : N]$.

The two new conditions are $N e\, G$ and $N r\, G$ for a normal subgroup of a group G. Note that for a saturated formation $\mathfrak{F}$, that $G_{\mathfrak{F}}\, e\, G$ and $G_{\mathfrak{F}}\, r\, G$.

Consider a solvable group G and Sylow system Σ for G. Let N be a normal subgroup of G such that $N e\, G$, and let π be the set of primes dividing $[G : N]$. By work done by Wright [110], he shows that $D = D_{N,\Sigma,\pi}$ covers all p-chief factors for $p \in \pi$ that are centralized by ${}^pN$ and avoids the remaining chief factors. Let

$$\{1\} = N_0 \lhd \cdots \lhd N_i = N \lhd N_{i+1} \lhd \cdots \lhd N_t = G$$

be a chief series for G. By definition of $N e\, G$, no p-chief factor of G in N is centralized by ${}^pN$. Thus, all chief factors N_j/N_{j-1}, for $1 \le j \le i$, are avoided by D.

Now consider a chief factor N_j/N_{j-1}, for $i + 1 \le j \le t$. Since ${}^{p}N/N = O_{p'}(G/N)$, N_j/N_{j-1} will be centralized by ${}^{p}N$. Thus, N_j/N_{j-1} is covered by D. Thus, by Lemma 5.1, G splits over N by D. This leads to the following result.

Theorem 7.16. [Proposition 1.4 in Wright [111]] *Let $N \trianglelefteq G$ for a solvable group G. Then, $Ne\, G$ if and only if $D_{H,\Sigma}$ complements N. In particular, if $Ne\, G$, then G splits over N.*

The next set of results by Wright require the following theorem.

Theorem 7.17. [Theorem 2.1 in Wright [111]] *Let $Nr_p\, G$ for G a solvable group, and let P be a Sylow p-subgroup of N. Suppose N is abelian. Then, ${}^{p}N$ centralizes no p-chief factor of G below N.*

Wright's Proof. Let G be a counterexample of minimal order, and let L be a minimal normal subgroup of G such that $L \le {}^{p}N$. Then, ${}^{p}N$ centralizes no p-chief factors between L and LN. If $|L| = q^{\alpha}$ for a prime q with $q \ne p$, then G is not a minimal counter example. Thus, L is a p-group where $L \le N \cap O_p(G) \le P$ and ${}^{p}N$ centralizes L. This implies $L \le P \cap Z({}^{p}N)$.

By Theorems 7.6 and 7.7, $P \cap Z({}^{p}N) \cap {}^{p}N_{\mathfrak{N}} = \{1\}$ or $L \cap {}^{p}N_{\mathfrak{N}} = \{1\}$. This implies that ${}^{p}N_{\mathfrak{N}} = \{1\}$ or that ${}^{p}N$ is nilpotent. Consequently, ${}^{p}N$ centralizes all chief factors of G in N. Given that $Nr_p\, G$, this implies $N = \{1\}$ and that G is not a minimal counter example. □

Let G be a solvable group, and let N be a normal subgroup of G with abelian Sylow p-subgroups for each prime p dividing $[G : N]$. By Theorem 7.17, ${}^{p}N$ centralizes no p-chief factors of G below N. This implies by Definition 7.15 that $Ne\, G$ or that G splits over N.

Corollary 7.18. [Corollary 2.2 in Wright [111]] *If $Nr\, G$, for G a solvable group, and if N has abelian Sylow subgroups for the primes dividing $[G : N]$. Then, $Ne\, G$ (and hence G splits over H).*

Theorem 7.17 can also be applied to obtain the following result.

Corollary 7.19. [Corollary 2.3 in Wright [111]] *Let $\mathfrak{F}$ be a locally induced formation with support π and let $L \trianglelefteq G$ for G a solvable group. Suppose that L contains a Hall π'-subgroup of G and that $L_{\mathfrak{F}}$ has abelian Sylow subgroups for the primes dividing $[G : L_{\mathfrak{F}}]$. Then, $L_{\mathfrak{F}}\, e\, G$ (and hence G splits over $L_{\mathfrak{F}}$).*

Applying the techniques used above, Wright also gets the following result.

Theorem 7.20. [Theorem 2.6 in Wright [111]] *Let $N \trianglelefteq G$ for G a solvable group. Suppose $G = XN$ with $X \cap N = N'$ and that N has abelian Sylow subgroups for the primes dividing $[G : N]$. Then, N has a complement in G.*

Wright also developed in [111] a variety of conditions for when all of the complements of a normal subgroup are conjugate.

Theorem 7.21. [Theorem 3.1 in Wright [111]] *Let $N \trianglelefteq G$, for G a solvable group, and suppose that $X \leq G$ with $NX = G$ and $N \cap X = \{1\}$. Suppose that for each prime p dividing $[G : N]$ the subgroup $O_{p'}(X)$ acts without a fixed point on each p-chief factor of G below N. Then, all complements to H in G are conjugate.*

Wright's Proof. If $N = \{1\}$, the result is trivial. Assume $N \neq \{1\}$ and that G is a minimal counterexample. Let Y be a subgroup of G such that $G = [N]Y$, and let L be a minimal normal subgroup of G such that $L \leq N$. By Lemma 1.11, $G/L = [N/L]LX/L = [N/L]LY/L$. Thus, LX/L and LY/Y are conjugate in G/L, which implies by Lemma 1.14 that LX and LY are conjugate in G. Given that any conjugate of Y is also a complement to N (Corollary 1.13), assume $T = LX = LY$.

Since L is a minimal normal subgroup of G, $L \leq O_p(G)$ for some prime p. If X and Y are Hall p'-subgroups of T, then they are conjugate by Theorem 2.21. Thus, X and Y are not Hall p'-subgroups of T and p divides $|X| = [G : N]$. Thus, by Maschke's Theorem (see A.11.4 in [34]), L is the direct sum of irreducible $O_{p'}(X)$-modules where $O_{p'}(X)$ acts fixed point freely on each. This means X and Y are conjugate in T, and consequently conjugate on G. This contradiction implies $T = LX = LY = G$ and that X and Y are maximal subgroups of G.

Given that $O_{p'}(G/L) = O_{p'}(X)L/L = O_{p'}(Y)L/L$, it follows that $O_{p'}(X)L = O_{p'}(Y)L$ with $O_{p'}(X) \cap L = O_{p'}(Y) \cap L = \{1\}$. This implies that $O_{p'}(X)$ and $O_{p'}(Y)$ are conjugate. Given that $L \nleq \mathcal{N}_G(O_{p'}(X))$ and $L \nleq \mathcal{N}_G(O_{p'}(Y))$, this implies $X = \mathcal{N}_G(O_{p'}(X))$ and $Y = \mathcal{N}_G(O_{p'}(Y))$. Thus, X and Y are conjugate in G, the final contradiction. □

Now let N be a normal subgroup of a solvable group G, with a normal Hall p'-subgroup, such that $N e G$. By Theorem 7.16, G splits over N and $G = [N]X$ for some subgroup X of G. Let p divide $[G : N]$. By Definition 7.15, ${}^pN$ centralizes no p-chief factor of G below N. Thus, ${}^pN$ acts fixed point freely on all p-chief factors of G below N. However, given that N has a normal Hall p'-subgroup, N will centralize each p-chief factor of G below N. Given that ${}^pN/N = O_{p'}(G/N)$ with $G/N \cong X$, ${}^pN = NO_{p'}(X)$ and applying Theorem 7.21, the following result is obtained.

Corollary 7.22. [Corollary 3.2 in Wright [111]] *Suppose that $N e G$, for a solvable group G, and that N has a normal Hall p'-subgroup for every prime p dividing $[G : N]$. Then, all complements to N in G are conjugate.*

Apply Corollary 7.18 and Corollary 7.22 to obtain the following result.

Corollary 7.23. [Corollary 3.3 in Wright [111]] *If $N\ r\ G$, for G a solvable group, and if N is abelian, then $N\ e\ G$ and all complements to N in G are conjugate.*

Wright also proves the following result. The proof is lengthy and is omitted here.

Theorem 7.24. [Theorem 3.4 in Wright [111]] *Let $N e G$, for G a solvable group, and let $D = D_{N,\Sigma}$. Suppose that for each prime p dividing $[G : N]$ the Sylow p-subgroups of D act fixed-point-freely on all p'-chief factors of G below N. Suppose, too, that D has quaternion-free Sylow 2-subgroups. Then, all complements to N in G are conjugate.*

This chapter concludes with a result by Bechtell. First a definition is needed.

Definition 7.25. Let G be a group, and let $\mathfrak{F}$ be a formation. Let $F_0(G) = G$, $F_1(G) = G_{\mathfrak{F}}$ and $F_i(G) = (F_{i-1}(G))_{\mathfrak{F}}$ for $i \geq 2$. If this terminates ($F_n(G) = \{1\}$) for some $n \geq 1$) then

$$G = F_0 > F_1 > \cdots > F_n = \{1\},$$

where $F_i = F_i(G)$, is a **$\mathfrak{F}$-derived series** of **$\mathfrak{F}$-derived length n.**

For example, if $\mathfrak{F} = \mathfrak{A}$, the formation of abelian groups, then the $\mathfrak{A}$-derived series is the derived series of a group. If $\mathfrak{F} = \mathfrak{N}$, the formation of nilpotent groups, then the $\mathfrak{N}$-derived series is the lower nilpotent series.

Using $\mathfrak{F}$-derived series, Bechtell [8] proves a slight variation of Carter's theorem 5.16. A simple lemma is need, which is presented without proof.

Lemma 7.26. [Lemma 2.4 in Bechtell [8]] *If a solvable group G having $\mathfrak{F}$-derived length $n + 1$, with respect to a normal formation $\mathfrak{F}$, splits over F_n and F_n is a minimal normal subgroup of G, then the complements of F_n are conjugate in G.*

Theorem 7.27. [Theorem 2.5 in Bechtell [8]] *If a solvable group G having $\mathfrak{F}$-derived length $n + 1$ with respect to a normal formation $\mathfrak{F}$ splits over F_n and F_n is a completely G-reducible abelian subgroup of G, then the complements of F_n are conjugate in G.*

Bechtell's Proof. Proceed by induction on the order of G. Let $F_n = N_1 \times \cdots \times N_k$, where $k \geq 1$ and where each N_i, for $1 \leq i \leq k$, is minimal normal in G. If $k = 1$, the result follows from Lemma 7.26.

Suppose $k \geq 2$. By Lemma 1.93, it follows that $F_n(G/N_1) = F_n/N_1$. Let H and K be complements to F_n in G. By Lemma 1.11,

$$G/N_1 = [F_n/N_1]N_1H/N_1 = [F_n/N_1]N_1K/N_1.$$

Now consider the subgroup $L = (N_2 \times \ldots \times N_k)H$ of G. It follows that $F_n(L) = N_2 \times \cdots \times N_k$. Given that F_n being completely G-reducible is the same as F_n being completely H-reducible, $N_1 \times \cdots \times N_k/N_1$ is completely G/N_1-reducible.

Thus, by induction, there exists an element $g \in G$ such that $(N_1K)^g = N_1H$ or that $N_1K^g = N_1H$.

Now consider $F_n(N_1H)$. It follows that $F_n(N_1H) \leq N_1$. Suppose that $F_n(N_1H) < N_1$ and let $N = F_n(N_1H)$. This implies

$$F_n(G) \leq N_2 \times \cdots \times N_k \times N < F_n,$$

which is a contradiction. Thus, $F_n(N_1H) = N_1$ with N_1 a minimal normal subgroup of N_1H. By Lemma 7.26, there exists an element $h \in N_1H$ such that $H = (K^g)^h = K^{gh}$. □

8 Groups with specific classes of subgroups complemented

Properties of a group G can be determined when specific classes of subgroups of G are complemented in G. One of the more famous results along this line is the one by Hall [52] stating that a group G is solvable if and only if every Sylow p-subgroup is complemented in G. A number of results along this line will be established in this chapter. However, an understanding of $\mathcal{K}$-groups is necessary before this investigation begins.

8.1 $\mathcal{K}$-groups

Definition 8.1. A group G is a **$\mathcal{K}$-group** if for every subgroup H of G there exists a subgroup K of G, such that $G = \langle H, K\rangle$ and $H \cap K = \{1\}$.

Note that for subgroups H and K from Definition 8.1, HK need not necessary be a subgroup of G. Now, $\mathcal{K}$-groups, which have trivial Frattini subgroups, were studied by Bechtell [5], Gross [47], Suzuki [102], and Zacher [113]. Subgroups of $\mathcal{K}$-groups are not necessarily $\mathcal{K}$-groups as demonstrated by the group S_4 (consider the Sylow 2-subgroup of S_4). It is not known if normal subgroups of $\mathcal{K}$-groups are $\mathcal{K}$-groups, but it will be shown that normal subgroups of solvable $\mathcal{K}$-groups are $\mathcal{K}$-groups. This first lemma concerning $\mathcal{K}$-groups stems from the fact that $\mathcal{K}$-groups have trivial Frattini subgroups.

Lemma 8.2. *A nilpotent group is a $\mathcal{K}$-group if and only if it is elementary abelian.*

Theorem 8.3. [Theorema in Zacher [113]] *A group G is a $\mathcal{K}$-group if and only if $\mathcal{F}(G)$ is the direct product of minimal normal abelian subgroups of G and $\mathcal{F}(G)$ has a complement which is a $\mathcal{K}$-group.*

Suzuki's Proof from [102]. ($\Rightarrow$) Suppose G is a $\mathcal{K}$-group, and consider $\mathcal{F}(G)$. Given that $\Phi(G) = \{1\}$, Theorem 1.69 implies $\mathcal{F}(G)$ is the direct product of abelian minimal normal subgroups of G. Furthermore, there exists a subgroup K of G, such that $G = \langle\mathcal{F}(G), K\rangle$ with $\mathcal{F}(G) \cap K = \{1\}$. Given that $\mathcal{F}(G) \trianglelefteq G$, this implies $G = [\mathcal{F}(G)]K$.

Consider $G/\mathcal{F}(G)$ and let $H/\mathcal{F}(G)$ be a subgroup of $G/\mathcal{F}(G)$. Thus, $H \leq G$ and there exists a subgroup L of G, such that $G = \langle H, L\rangle$ with $H \cap L = \{1\}$. Consequently, by the modular identity (Lemma 1.6),

$$\mathcal{F}(G)L \cap H = \mathcal{F}(G)(H \cap L) = \mathcal{F}(G).$$

DOI 10.1515/9783110480214-008

Since $\langle H, \mathcal{F}(G)L\rangle = G$, it follows that $G/\mathcal{F}(G) = \langle H/\mathcal{F}(G), L\mathcal{F}(G)/\mathcal{F}(G)\rangle$ and $G/\mathcal{F}(G)$ is a $\mathcal{K}$-group. Given that $K \cong G/\mathcal{F}(G)$, K is a $\mathcal{K}$-group.

($\Leftarrow$) Suppose $G = [\mathcal{F}(G)]K$ for $K \leq G$, with K a $\mathcal{K}$-group and $\mathcal{F}(G)$ the direct product of minimal normal abelian subgroups of G. Let $H \leq G$. Then, $H\mathcal{F}(G)/\mathcal{F}(G) \leq G/\mathcal{F}(G)$. Since $G/\mathcal{F}(G)$ is a $\mathcal{K}$-group, there exists a subgroup $L/\mathcal{F}(G)$ of $G/\mathcal{F}(G)$, such that $G/\mathcal{F}(G) = \langle H\mathcal{F}(G)/\mathcal{F}(G), L/\mathcal{F}(G)\rangle$ with $H\mathcal{F}(G) \cap L = \mathcal{F}(G)$. This implies $G = \langle H\mathcal{F}(G), L\rangle$.

By the modular identity (Lemma 1.6),

$$\mathcal{F}(G) = \mathcal{F}(G)H \cap L = \mathcal{F}(G)(H \cap L),$$

which implies $H \cap L \leq \mathcal{F}(G)$ or that

$$H \cap L = H \cap L \cap \mathcal{F}(G) = H \cap \mathcal{F}(G).$$

Let N be maximal, such that $N \trianglelefteq G$, $N \leq \mathcal{F}(G)$, and $N \cap H = \{1\}$. Consider $T = N(L \cap K)$. By the modular identity (Lemma 1.6),

$$\begin{aligned} H \cap T &= H \cap N(L \cap K) \\ &= H \cap (NK \cap L) \\ &= H \cap L \cap NK \\ &= H \cap \mathcal{F}(G) \cap NK \\ &= H \cap N(\mathcal{F}(G) \cap K) \\ &= H \cap N \\ &= \{1\}. \end{aligned}$$

Now, consider $\langle H, T\rangle$.

Case 1) $\mathcal{F}(G) \leq \langle H, T\rangle$. In this case, it readily follows that

$$\begin{aligned} \langle H, T\rangle &= \langle H, T\mathcal{F}(G)\rangle \\ &= \langle H, \mathcal{F}(G)N(L \cap K)\rangle \\ &= \langle H, \mathcal{F}(G)(L \cap K)\rangle \\ &= \langle H, \mathcal{F}(G)K \cap L\rangle \\ &= \langle H, L\rangle \\ &= G. \end{aligned}$$

Case 2) $\mathcal{F}(G) \nleq \langle H, T\rangle$. Let M be a maximal subgroup of G, such that $\langle H, T\rangle \leq M$. Since $G = \langle H, T, \mathcal{F}(G)\rangle$, $\mathcal{F}(G) \nleq M$. Thus, there exists a minimal normal subgroup R of G, such that $R \leq \mathcal{F}(G)$ and $R \nleq M$. This implies $G = RM$ with $R \cap M = \{1\}$. Now, $NR \trianglelefteq G$, such that $NR \leq \mathcal{F}(G)$ with $NR \cap H = \{1\}$ (if $NR \cap H \neq \{1\}$, then $nr = h$, for $n \in N$, $r \in R$, and $h \in H$, which implies $r = n^{-1}h$ or that $r \in NT$, a contradiction). This contradicts the fact that N is maximal with $N \trianglelefteq G$, $N \leq \mathcal{F}(G)$, and $N \cap H = \{1\}$. Consequently, $G = \langle H, T\rangle$ with $H \cap T = \{1\}$. □

In general, $\mathcal{K}$-groups are not solvable as demonstrated by the simple group $\mathcal{A}_5$. In fact, Costantini and Zacher [31] proved the following result (the proof is omitted).

Theorem 8.4. [Theorem in Costantini and Zacher [31]] *Every simple group is a $\mathcal{K}$-group.*

Before classifying solvable $\mathcal{K}$-groups, the following result is needed.

Lemma 8.5. *Every homomorphic image of a solvable $\mathcal{K}$-group is a $\mathcal{K}$-group.*

Beweis. Let G be a solvable $\mathcal{K}$-group, and let $N \trianglelefteq G$ with $H/N \leq G/N$. Thus, $H \leq G$ and there exists a subgroup L of G, such that $G = \langle H, L\rangle$ with $H \cap L = \{1\}$. Since G is solvable, there exists a minimal normal abelian subgroup M of G, such that $M \leq N$. By Corollary 3.2, there exists a subgroup T of G, such that $G = [M]T$.

Consider the subgroup $M(NL \cap T)$ of G. By the modular identity (Lemma 1.6), $M(NL \cap T) = MT \cap NL = NL$, which indicates that $N \leq M(NL \cap T)$. Furthermore, a simple calculation shows

$$H \cap M(NL \cap T) = H \cap MT \cap NL = H \cap NL = N(H \cap L) = N,$$

with

$$\langle H, M(NL \cap T)\rangle = \langle H, MT \cap NL\rangle = \langle H, NL\rangle = G.$$

Thus, $M(NL \cap T)/N \leq G/N$ with $G/N = \langle H/N, M(NL \cap T)/N\rangle$ and $H/N \cap M(NL \cap T)/N = \{1_{G/N}\}$. Thus, G/N is a $\mathcal{K}$-group. □

Theorem 8.6. [Main Result in Zacher [113]] *A solvable group G is a $\mathcal{K}$-group if and only if G contains a normal series*

$$\{1\} = N_0 \leq N_1 \leq \cdots \leq N_r = G,$$

such that N_{i+1}/N_i is the maximal nilpotent normal subgroup of G/N_i and $\Phi(G/N_i) = \{1_{G/N_i}\}$ for $i = 0, \ldots, r-1$.

Suzuki's Proof from [102]. (⇒) First, recall that a $\mathcal{K}$-group has a trivial Frattini subgroup. Let $N_0 = \{1\}$. For $i \geq 1$, let $N_i = \mathcal{F}(G/N_{i-1})$. Since G is solvable, there exists an $r \geq 1$, such that $N_r = G$. Furthermore, as indicated by Lemma 8.5, G/N_i is a $\mathcal{K}$-group for $0 \leq i \leq r-1$, forcing $\Phi(G/N_i) = \{1_{G/N_i}\}$.

($\Leftarrow$) Suppose G has a normal series

$$\{1\} = N_0 \leq N_1 \leq \cdots \leq N_r = G,$$

such that N_{i+1}/N_i is the maximal nilpotent normal subgroup of G/N_i and $\Phi(G/N_i) = \{1_{G/N_i}\}$ for $i = 0, \ldots, r-1$. Proceed by induction on r, and consider $N_1 = \mathcal{F}(G)$. Given that $\Phi(G) = \{1\}$, it follows by Theorem 1.69 that N_1 is the direct product of the minimal normal abelian subgroups of G.

To show that N_1 has a complement in G, let $H \leq N_1$ be maximal, such that there exists a subgroup K of G satisfying $G = \langle H, K\rangle$ and $H \cap K = \{1\}$. Suppose $H < N_1$. This implies $N_1 \cap K \neq \{1\}$ with $N_1 \cap K \lhd G$. Let N be a minimal normal subgroup of G with $N \leq N_1 \cap K$. By Corollary 3.2, $G = [N]M$ for some maximal subgroup M of G. It then follows from the modular identity (Lemma 1.6) that

$$NH \cap M \cap K = NH \cap K \cap M = N(H \cap K) \cap M = N \cap M = \{1\}$$

and

$$\langle NH, M \cap K\rangle = \langle H, N(M \cap K)\rangle = \langle N, NM \cap K\rangle = \langle H, K\rangle = G.$$

This contradicts the maximality of H. Therefore, $H = N_1$ and, given that $G = \langle N, K\rangle$ with $N \cap K = \{1\}$, it follows that $G = [N_1]K$. By induction, $G/N_1 \cong K$ is a $\mathcal{K}$-group. Thus, by Theorem 8.3, G is a $\mathcal{K}$-group. □

At this point, a fact stated earlier can be proven.

Lemma 8.7. [Gross [47]] *Every normal subgroup of a solvable $\mathcal{K}$-group is a $\mathcal{K}$-group.*

Beweis. Let G be a solvable $\mathcal{K}$-group, and let $N \trianglelefteq G$. Since N is solvable (Lemma 1.25), $N_1 = \mathcal{F}(N) \neq \{1\}$. Corollary 1.54 implies $\Phi(N) \leq \Phi(G)$ or, since $\Phi(G) = \{1\}$, that $\Phi(N) = \{1\}$. Thus, by Theorem 1.69, N_1 is the direct product of the minimal normal subgroups of N.

Given that N_1 is characteristic in N, N_1 is normal in G. By Lemma 8.5, G/N_1 is a $\mathcal{K}$-group. This implies $\Phi(G/N_1) = \{1_{G/N_1}\}$ and that $\Phi(N/N_1) = \{1_{N/N_1}\}$ as $\Phi(N/N_1) \leq \Phi(G/N_1)$ by Corollary 1.54. Continuing in this manner, N contains a normal series

$$\{1\} = N_0 \leq N_1 \leq \cdots \leq N_r = N,$$

such that such N_{i+1}/N_i is the maximal nilpotent normal subgroup of G/N_i and $\Phi(G/N_i) = \{1_{G/N_i}\}$ for $i = 0, 1, \ldots, r-1$. Thus, by Theorem 8.6, N is $\mathcal{K}$-group. □

It will be shown later that groups with elementary abelian Sylow subgroups are $\mathcal{K}$-groups. However, as demonstrated by $\mathcal{S}_4$, not all $\mathcal{K}$-groups have elementary abelian Sylow subgroups. To account for this possibility, Bechtell, corrected by Kotzen, proved the following result.

Theorem 8.8. [Theorem 1 in Bechtell [5], Theorem 1 in Kotzen [74]] *A group G is a solvable $\mathcal{K}$-group if and only if G is a subdirect product of a finite collection of solvable $\mathcal{K}$-groups H_i, where $1 \le i \le n$, such that each H_i is isomorphic to a subgroup of G and each H_i possess a unique minimal normal subgroup.*

Kotzen's Proof. ($\Rightarrow$) Let G be a solvable $\mathcal{K}$-group. Since G is solvable with $\Phi(G) = \{1\}$, G is the subdirect product of a finite collection of solvable groups H_i, for $1 \le i \le n$, such that each H_i has a unique minimal normal subgroup (see A.15.9 in [34]). Define $\rho_k : G \to H_k$, for some k, with $1 \le k \le n$, to be the projective mapping with kernel L. Thus, $L \lhd G$, and, since G is a $\mathcal{K}$-group, there exists a subgroup K of G, such that $G = \langle L, K\rangle = LK$ with $L \cap K = \{1\}$. Since $G/L \cong H_k$. it follows that $K \cong H_k$. Thus, by Lemma 8.5, H_k is a $\mathcal{K}$-group.

($\Leftarrow$) Suppose G is a subdirect product of a finite collection of solvable $\mathcal{K}$-groups H_i, where $1 \le i \le n$, such that each H_i is isomorphic to a subgroup of G and each H_i possess a unique minimal normal subgroup. Proceed by induction on the order of G. For $1 \le i \le n$, let $\rho_i : G \to H_i$ denote the projective from G into H_i with kernel L_i. It follows that $\bigcap_{i=1}^n L_i = \{1\}$. Since each H_i is a $\mathcal{K}$-group, $\Phi(H_i) = \{1\}$. By Lemma 1.52, it follows that $\rho_i(\Phi(G)) \le \Phi(H_i) = \{1\}$ or that $\Phi(G) \le \bigcap_{i=1}^n L_i$. This means $\Phi(G) = \{1\}$.

Given that G is solvable, $\mathcal{F}(G) \ne \{1\}$. Since $\Phi(G) = \{1\}$, $\mathcal{F}(G)$ is elementary abelian by Theorem 1.69. Given that $\mathcal{F}(G)$ is abelian with $\mathcal{F}(G) \cap \Phi(G) = \{1\}$, it follows by Theorem 3.9 that $G = [\mathcal{F}(G)]K$ for some subgroup K of G.

Recall that H_i is a solvable $\mathcal{K}$-group with a unique minimal normal subgroup. This implies that $\mathcal{F}(H_i)$ must be the unique minimal normal subgroup of H_i. Now, note that for each i, $H_i = \rho_i(\mathcal{F}(G))\rho_i(K)$. There are two cases to consider.

Case 1) $\rho_i(\mathcal{F}(G)) \cap \rho_i(K) = \{1\}$. In this case, $H_i/\rho_i(\mathcal{F}(G)) \cong \rho_i(K)$, and by Lemma 8.5, $\rho_i(K)$ is a solvable $\mathcal{K}$-group.

Case 2) $\rho_i(\mathcal{F}(G)) \cap \rho_i(K) = M_i \ne \{1\}$. Here $M_i \le \rho_i(\mathcal{F}(G)) \le \mathcal{F}(H_i)$. Given that M_i is abelian with $H_i = \rho_i(\mathcal{F}(G))\rho_i(K)$, M_i is normal in H_i, which forces $M_i = \mathcal{F}(H_i)$. This means $\rho_i(\mathcal{F}(G)) \le \rho_i(K)$ or that $\rho_i(K) = H_i$, which is a solvable $\mathcal{K}$-group.

Thus, for each i, $1 \le i \le n$, $\rho_i(K)$ is a solvable $\mathcal{K}$-group. Furthermore, if $\rho_i(K) \ne H_i$, then $\rho_i(K)$ is a subdirect product of solvable $\mathcal{K}$-groups. This means K is a subdirect product of solvable $\mathcal{K}$-groups, such that each is isomorphic to a subgroup of K and each has a unique minimal normal subgroup. By induction, K is a solvable $\mathcal{K}$-group. Therefore, by Theorem 8.3, G is a solvable $\mathcal{K}$-group. □

Now, suppose G is a supersolvable $\mathcal{K}$-group. If G is a p-group, for some prime p, then G is elementary abelian as $\Phi(G) = \{1\}$. Assume otherwise and let q be the largest prime dividing the order of G. By Theorem 1.43, the Sylow q-subgroup Q of G is normal in G. It follows from Corollary 1.54 that $\Phi(Q) \le \Phi(G) = \{1\}$, which implies that Q is elementary abelian. By Theorem 1.42 and Lemma 8.5, G/Q is a supersolvable $\mathcal{K}$-group. By induction, G/Q has elementary abelian Sylow subgroups. Thus, G has elementary abelian Sylow subgroups. This motivates the following result by Suzuki. Its proof is similar to the proof of Theorem 8.22 and is omitted.

Theorem 8.9. [Theorem I.24 in Suzuki [102]] *The following properties of a group G are equivalent.*
(i) *G is a supersolvable $\mathcal{K}$-group;*
(ii) *G is isomorphic with a subgroup of a direct product of groups with square-free order;*
(iii) *every subgroup H of G has a complement K in G.*

In a $\mathcal{K}$-group G, it follows that $\Phi(N) = \{1\}$ for all normal subgroups in G. This motivates the examination of groups in which the Frattini subgroup of every subgroup is trivial.

Definition 8.10. A group G is an **elementary** group if $\Phi(H) = \{1\}$ for all $H \leq G$.

Elementary groups were introduced and studied by Bechtell [4]. It is trivial to show that every subgroup and homomorphic image of an elementary group is an elementary group.

Lemma 8.11. [Lemma 2.1 in Bechtell [4]] *An elementary group is a $\mathcal{K}$-group.*

Bechtell's Proof. Let H be a proper nontrivial subgroup of G. Since $\Phi(G) = \{1\}$, there is a maximal subgroup M of G, such that $G = \langle H, M\rangle$. Suppose $H \cap M \neq \{1\}$. Since $\Phi(M) = \{1\}$, $M = \langle H \cap M, M_1\rangle$ for M_1 maximal in M. This implies $G = \langle H, H \cap M, M_1\rangle = \langle H, M_1\rangle$. Continue until subgroup M_n is obtain, for $n \geq 1$, such that $G = \langle H, M_n\rangle$ with $H \cap M_n = \{1\}$. This implies G is a $\mathcal{K}$-group. □

The next result follows directly.

Corollary 8.12. [Corollary 2.1 in Bechtell [4]] *An elementary group splits over each normal subgroup.*

The converse of Corollary 8.12 is not true as the symmetric group $\mathcal{S}_4$ indicates.

Theorem 8.13. [Theorem 2.2 in Bechtell [4]] *A group G is an elementary group if and only if*
(i) *G is a $\mathcal{K}$-group and*
(ii) *each subgroup of G is a $\mathcal{K}$-group.*

Beweis. The conditions are sufficient by Lemma 8.11. For the converse, consider group G that satisfies condition (i) and (ii). Let H be a subgroup of G. Since H is a $\mathcal{K}$-group, $\Phi(H) = \{1\}$. Thus, G is an elementary group. □

8.2 *xD*-groups

To put the rest of the results presented in this chapter in context, a comprehensive definition is needed.

Definition 8.14. A group G is an ***xP*-group** if every nontrivial x-subgroup satisfies condition P, where x and P can have the following values.

$$\begin{aligned} x &= a \text{ (arbitrary subgroup);} \\ &= n \text{ (normal subgroup);} \\ &= c \text{ (characteristic subgroup);} \\ &= f \text{ (fully invariant subgroup);} \end{aligned}$$

$$\begin{aligned} P &= D \text{ (is a direct factor);} \\ &= C \text{ (has a complement);} \\ &= S \text{ (has a proper supplement);} \\ &= PNS \text{ (has a proper normal supplement);} \\ &= CS \text{ (has a proper characteristic supplement).} \end{aligned}$$

Given the scope of this book, this chapter will focus on *xD*- and *xC*-groups for $x = a, n,$ and c. Information on *xS*-, *xPNS*-, and *xCS*-groups can be found in [69].

Theorem 8.15. [Theorem in Kertész [70]] *A group G is an aD-group if and only if G is abelian and the order of each element of G is a square-free number.*

Beweis. ($\Rightarrow$) Let G be an *aD*-group. Then, every subgroup of G is normal in G and G is either abelian or $G = Q_8 \times A_2 \times A$, where Q_8 is the quaternion group of 8 elements, A_2 is an elementary abelian 2-group, and A is an abelian group of odd order (see 5.3.7 in [87]).

Assume that G is nonabelian. Now, Q_8 is normal in G, so by Corollary 1.54 $\Phi(Q_8) \leq \Phi(G)$. This means $\Phi(G) \neq \{1\}$. Thus, $G = \Phi(G) \times K$ for some subgroup K of G, which implies

$$G = \langle \Phi(G), K \rangle = \langle K \rangle = K,$$

a contradiction. Thus, G is abelian.

Let $x \in G$ and suppose p^2 divides $|x|$. Then, G contains a normal cyclic subgroup of order p^2, which will have a nontrivial Frattini subgroup. A similar argument to the one presented in the previous paragraph results in a contradiction. Thus, G is abelian and each element is of square-free order.

(⇐) Let G be an abelian group with each element of square-free order, and let H be a nontrivial subgroup of G. By Lemma 1.60, there exists a subgroup K of G, such that $G = HK$ with $H \cap K \leq \Phi(K)$.

Suppose $\Phi(K) \neq \{1\}$. Then, p divides $|\Phi(K)|$ for some prime p. It then follows that p divides $|K/\Phi(K)|$ by Lemma 1.51. Given that K is abelian, Lemma 1.55 implies that K has an element of order p^n, for some $n \geq 2$. This is a contradiction. Thus, $\Phi(K) = \{1\}$ and $G = HK$ with $H \cap K = \{1\}$. Given that G is abelian, it follows that $G = H \times K$. □

Theorem 8.15 classifies *aD*-groups. The class of *nD*-groups was classified by Wiegold [108]. Before stating his result, a simple lemma is needed.

Lemma 8.16. *Every normal subgroup of an nD-group is an nD-group.*

Beweis. Let G be an *nD*-group, let N be a nontrivial normal subgroup of G, and let M be a nontrivial normal subgroup of N. Since N is direct factor of G, M is normal in G. Thus, $G = M \times H$ for some subgroup H of G. Thus, by Lemma 1.11, $N = M \times (N \cap H)$ and M is a direct factor of N. □

Theorem 8.17. [Corollary 4.2, Theorem 4.4, and Theorem 4.5 in Wiegold [108]] *A group G is an nD-group if and only if G is a direct product of simple groups.*

Wiegold's Proof. (⇒) Let G be an *nD*-group and proceed by induction on the order of G. If G is simple, the result follows. Suppose G is not simple, and let N be a minimal normal subgroup of G. Then, $G = N \times H$ for some subgroup H of G. First, note that this means that N must be simple. Furthermore, by Lemma 8.16, H is an *nD*-group. Thus, by induction, H is the direct product of simple groups. As a result, G is a direct product of simple groups.

(⇐) Let G be a direct product of simple groups. Then,

$$G = G_1 \times \cdots \times G_n \times A_1 \times \cdots \times A_m,$$

where each G_i, for $1 \leq i \leq n$, is a nonabelian simple group and each A_j, for $1 \leq j \leq m$, is an abelian simple group. It follows directly that $G = G' \times A$, where $G' = G_1 \times \cdots \times G_n$ and $A = Z(G) = A_1 \times \cdots \times A_m$.

First, it will be shown that G' is an *nD*-group. Let N be a normal subgroup of G', and let $x \in N$. It follows for all i, where $1 \leq i \leq n$, that $G_i \cap N = \{1\}$ or that $G_i \cap N = G_i$. Consider the case that for some k, where $1 \leq k \leq n$, that $G_k \cap N = \{1\}$. Suppose that $x = g_1 \cdots g_n$, where $g_i \in G_i$ with $g_k \neq 1$. Given that $Z(G_k) = \{1\}$, there exists an element $h \in G_k$, such that $[g_k, h] \neq 1$ with $[g_k, h] \in G_k$. Now, consider the element $[x, h]$ which is in N. A simple calculation shows

$$\begin{aligned}[x, h] &= x^{-1}h^{-1}xh \\ &= g_1^{-1} \cdots g_k^{-1} \cdots g_n^{-1} g_1^h \cdots g_k^h \cdots g_n^h\end{aligned}$$

$$= g_k^{-1} g_k^h$$
$$= [g_k, h].$$

This is a contradiction as $G_k \cap N = \{1\}$. This means that if $G_l \cap N = \{1\}$ for any l, with $1 \leq l \leq n$, then $x = g_1 \cdots g_n$, where $g_i \in G_i$, must satisfy $g_l = 1$.

Let K be a subgroup of G, such that $K = \langle G_i | N \cap G_i = G_i \rangle$. The work done in the previous paragraph implies $N \leq K$. However, given that $G_i \leq N$ when $N \cap G_i = G_i$, it also follows that $K \leq N$. This means $N = K$ and that N is a direct factor of G'.

To show that G is an *nD*-group, let M be a normal subgroup of G. Given that G' is an *nD*-group with $M \cap G'$ normal in G', it follows that $G' = (M \cap G') \times D$ for some subgroup D of G'. Furthermore, by Theorem 8.15, A is an *nD*-group. This means $A = (M \cap A) \times B$, where B is a subgroup of A. Note that $(M \cap G') \times (M \times A) \leq M$. Now, let $m \in M$. Then, $m = g_{r_1} \cdots g_{r_k} a_{t_1} \cdots a_{t_l}$ for $g_{r_i} \in G_{r_i}$ and $a_{t_j} \in A_{t_j}$. A similar argument to one presented earlier in this proof shows that $g_{r_1} \cdots g_{r_k} \in M \cap G'$. Since $a_{t_1} \cdots a_{t_l} \in M \cap A$, it follows that $M \leq (M \cap G') \times (M \cap A)$ or that $M = (M \cap G') \times (M \cap A)$. Consequently,

$$\begin{aligned} G &= G' \times A \\ &= (M \cap G') \times D \times (M \cap A) \times B \\ &= (M \cap G') \times (M \cap A) \times D \times B \\ &= M \times (D \times B) \end{aligned}$$

and M is a direct factor of G. □

Clearly, in terms of group classes, $aD \subseteq nD \subseteq cD$. Now, let S be your favorite nonabelian simple group. The group $G = S \times S$ shows that $aD \subset nD$. Christensen [28] showed that $nD = cD$. Before proving this result, a simple lemma is needed.

Lemma 8.18. *Every characteristic subgroup of a cD-group is a cD-group.*

Beweis. Let G be a *cD*-group, and let C be a characteristic subgroup of G. Now, let H be a characteristics subgroup of C. Then, H is characteristic in G and $G = H \times K$ for some subgroup K of G. By Lemma 1.11, it follows that $C = H \times (C \cap K)$. □

Theorem 8.19. [Theorem 3.1 in Christensen [28]] *Every cD-group is an nD-group.*

Christensen's Proof. Let G be a *cD*-group, and let M_1 be a maximal characteristic subgroup of G. Then, $G = M_1 \times K_1$, where $K_1 < G$ and K_1 has no characteristic subgroups. Let M_2 be a maximal characteristic subgroup of M_1. Since M_1 is a *cD*-group by Lemma 8.18, $M_1 = M_2 \times K_2$, where $K_2 \leq M_1$ and K_2 has no characteristic subgroups. Continue in this manner to obtain $G = K_n \times \cdots \times K_1$, where $n \geq 1$ and K_i, for $1 \leq i \leq n$ has no characteristic subgroups. This means that each K_i is a direct product of isomorphic simple groups (see I.9.12 in [66]). Thus, by Theorem 8.17, G is an nD-group. □

Christensen [28] uses Theorem 8.19 to prove that any solvable xD-group is abelian.

8.3 *aC*-groups

Hall's observation that a group G is solvable if and only if every Sylow p-subgroup is complemented in G motivated him to examine groups that satisfied a stronger complementation condition.

Definition 8.20. A group G is an ***aC*-group** or is **complemented** if for every subgroup H of G there exists a subgroup K in G, such that $G = HK$ with $H \cap K = \{1\}$.

Before examining aC-groups note that every aC-group is a $\mathcal{K}$-group, but as will be shown later, not every $\mathcal{K}$-group is an aC-group.

Lemma 8.21.
(i) *Any subgroup of an aC-group is an aC-group.*
(ii) *The direct product of two aC-groups is an aC-group.*
(iii) *A group of square-free order is an aC-group.*
(iv) *If G is an aC-group, then* $\Phi(G) = \{1\}$.

Beweis. For (i), let H be a subgroup of an aC-group G and let K be a subgroup of H. Then, K is a subgroup of G and $G = KL$, with $K \cap L = \{1\}$, for some subgroup L of G. It then follows by the modular identity (Lemma 1.6) that

$$K(L \cap H) = KL \cap H = H$$

and that K is complemented in H.

To prove (ii), let G_1 and G_2 be aC-groups and let H be a subgroup of $G_1 \times G_2$. Consider the subgroups $H_1 = \{x_1 \in G_1 | x_1 y_2 \in H \text{ for some } y_2 \in G_2\}$ and $H_2 = \{x_2 \in G_2 | y_1 x_2 \in H \text{ for some } y_1 \in G_1\}$ of G_1 and G_2 respectively. Since G_1 and G_2 are aC-groups, there exists subgroups $K_1 \leq G_1$ and $K_2 \leq G_2$, such that $G_1 = H_1 K_1$, with $H_1 \cap K_1 = \{1\}$ and $G_2 = H_2 K_2$, with $H_2 \cap K_2 = \{1\}$.

Consider the subgroup $K_1 \times K_2$ of $G_1 \times G_2$. Let $g \in H \cap (K_1 \times K_2)$. Then, $g = xy$, with $x \in G_1$ and $y \in G_2$, where $xy \in H$ and $xy \in K_1 \times K_2$. This implies $x \in H_1 \cap K_1 = \{1\}$ and $y \in H_2 \cap K_2 = \{1\}$ or that $H \cap (K_1 \times K_2) = \{1\}$.

Now, let $g = g_1 g_2 \in G$, where $g_1 \in G_1$ and $g_2 \in G_2$. Then, $g_1 = h_1 k_1$, with $h_1 \in H_1$ and $k_1 \in K_1$. Since $h_1 \in H_1$, there exists an element $t_2 \in G_2$, such that $h_1 t_2 \in H$. Thus, $t_2^{-1} g_2 \in G_2$ and $t_2^{-1} g_2 = h_2 k_2$ for $h_2 \in H_2$ and $k_2 \in K_2$. This implies

$$g = g_1 g_2 = h_1 k_1 t_2 h_2 k_2 = h_1 t_2 h_2 k_1 k_2,$$

where $h_1 t_2 h_2 \in H$ and $k_1 k_2 \in K_1 \times K_2$. Thus, H is complemented in $G_1 \times G_2$ by $K_1 \times K_2$.

For (iii), first note that any group G of square-free order has cyclic Sylow p-subgroups, implying that G is solvable (see p. 163-4 in [21]). Furthermore, any subgroup of G is a Hall π-subgroup and trivially has a complement.

To prove (iv), suppose that $\Phi(G) \neq \{1\}$. Then, there exists a proper subgroup K of G, such that $G = \Phi(G)K$. This implies $G = \langle \Phi(G), K \rangle = \langle K \rangle = K$, a contradiction. □

Complemented or aC-groups can now be classified.

Theorem 8.22. [Theorem 1 in Hall [53]] *A group G is an aC-group if and only if it is isomorphic to a subgroup of a direct product of groups of square-free order.*

Hall's Proof. The necessity of the condition follows from Lemma 8.21.

Now, let G be an aC-group, and let P be a Sylow p-subgroup of G. By Lemma 8.21, P is an aC-group with $\Phi(P) = \{1\}$. This implies that P is elementary abelian.

Given that each Sylow subgroup of G has a complement in G, it follows that G is solvable [52]. Let H/K be a chief factor of G of order p^α for some prime p and $\alpha \geq 1$. Suppose that $\alpha > 1$. Then, there exists a subgroup H_1 of G, such that $K < H_1 < H$. Thus, there is a subgroup L_1 of G that complements H_1 in G. Let $L_2 = H \cap L_1$, and let $L = KL_2$. The subgroup L satisfies $K < L < H$.

Since $H \trianglelefteq G$, it follows that $H \cap L_1 \trianglelefteq L_1$ or that $L_2 \trianglelefteq L_1$. Given that $L = KL_2$ with $K \triangleleft G$, L is normalized by L_1. Since $L \triangleleft H$, this implies $L \trianglelefteq HL_1$ or that $L \trianglelefteq G$. This contradiction implies $\alpha = \{1\}$. Thus, each chief factor of G is cyclic.

Now, consider a group G, such that G has elementary abelian Sylow p-subgroups and all chief factors cyclic. It will be shown that G is isomorphic to a subgroup of a direct product of groups of square-free order. Proceed by induction on the order of G. Let x be any nontrivial element of G, and let N_x be that largest normal subgroup of G, such that $x \notin N_x$. If no such N_x exists, then G has a unique minimal normal subgroup.

First, assume that N_x exists for each $x \in G$. Then, G/N_x is isomorphic to a subgroup of a direct product of groups of square-free order. Consequently, G is isomorphic to a subgroup of

$$\prod_{x \in G, x \neq 1} G/N_x,$$

and is therefore isomorphic to a subgroup of a direct product of groups of square-free order.

Continue assuming that G has a unique minimal normal subgroup M. Then, M is cyclic of order p, for p a prime, with $M = \langle y \rangle$ for $y \in G$. Let $N = C_G(\{y\})$. Then, N is normal in G with $M \leq N$. Suppose there exists a normal subgroup H of G, such that H/M is a chief factor of G with $H \leq N$. Thus, H/M is cyclical of order q, for q a prime. If $p \neq q$, then H contains a characteristic Sylow q-subgroup Q. Thus, $Q \triangleleft G$, which contradicts the fact that M is the unique minimal normal subgroup of G.

Suppose $p = q$. This means H is an elementary abelian p-group of order p^2 with G acting on H via conjugation. Thus, there is a linear representation L of G determined via the action of G on H. Since G has abelian Sylow p-subgroups, p does not divide the order of L. Thus, L is completely reducible (see 12.1.3 in [96]) and $H = \langle y\rangle \times \langle z\rangle$ with $\langle z\rangle \lhd G$. This contradicts the fact that M is the unique minimal normal subgroup of G.

This implies $N = \langle y\rangle$ and that $G/\langle y\rangle$ is isomorphic to a subgroup of the automorphism group of $\langle y\rangle$. Given that $Aut(\langle y\rangle)$ is cyclical of order $p - 1$ (see 5.7.11 in [96]), G is metacyclic of order pt, where t divides $p - 1$. Since all Sylow subgroups of G are elementary abelian, t must be of square-free order. Thus, G is a group of square-free order. □

The next result by Hall immediately follows.

Theorem 8.23. [Theorem 2 in Hall [53]] *A group G is an aC-group if and only if its Sylow subgroups are elementary abelian and its chief factors are cyclical.*

Clearly, complemented or aC-groups are solvable with each normal subgroup having a complement. However, if every normal subgroup of a group G has a complement in G, the group is not necessarily solvable. As an example, let G be the direct product of any two nonabelian simple groups. Another property of complemented groups is given after the following lemma.

Lemma 8.24. *For a group G of square-free order, G' and G/G' are cyclic.*

Beweis. Since an abelian group of square-free order is cyclic, it follows that G/G' is cyclic. Consider the subgroups G'' and G''' of G. It also follows that G'/G'' and G''/G''' are cyclic.

Proceed by assuming that $G''' = \{1\}$. Given that G'' is cyclic and $G/C_G(G'')$ is isomorphic to a subgroup of $Aut(G'')$, it follows that $G/C_G(G'')$ is abelian. This means $G' \leq C_G(G'')$, which implies $G'' \leq Z(G')$. Given that G'/G'' is cyclic, $G'/G'' = \langle gG''\rangle$ for some $g \in G'$. This means $G' = \langle G'', g\rangle$. Thus, G' is abelian, forcing $G'' = \{1\}$. Given that G' is abelian, it is cyclic. □

Corollary 8.25. *Let G be an aC-group. Then, G' is abelian and $G = [G']H$ for $H \leq G$ with H abelian.*

Beweis. By Theorem 8.22, $G = G_1 \times \cdots \times G_n$, where each G_i, for $1 \leq i \leq n$, is a group of square-free order. Since $G' = G_1' \times \cdots \times G_n'$, it follows that G' is abelian by Lemma 8.24. Since G is an aC-group, $G = [G']H$ for $H \leq G$. Given that $H \cong G/G'$, H is also abelian. □

Note that Corollary 8.25 implies that an aC-group is metabelian.

8.4 nC-groups

Now, nC-groups and cC-groups are a bit more complicated. First, note that as classes of groups $aC \subseteq nC \subseteq cC$. Any nonabelian simple group shows that $aC \subset nC$. It was proven independently by Dinerstein [32] and Hofmann [61] that as classes of groups $nC = cC$.

Theorem 8.26. [Theorem 1 in Dinerstein [32] and Main Theorem in Hofmann [61]] *A group G is an nC-group if and only if it is a cC-group.*

Now, to examine the structure of nC-groups.

Lemma 8.27. *Let G be an nC-group. Then, the following conditions hold:*
(i) $\Phi(G) = \{1\}$;
(ii) *every normal subgroup of G is an cC-group;*
(iii) *every homomorphic image of G is an nC-group;*
(iv) *the direct product of two nC-groups is an nC-group.*

Beweis. Condition (i) trivially follows. To prove (ii), let H be a normal subgroup of G, and let N be characteristic subgroup of H. Since N is normal in G, $G = [N]K$ for some subgroup K of G. It then follows by Lemma 1.11 that $H = [N](H \cap K)$ and that N has a complement in H.

For (iii), let $N \trianglelefteq G$ and consider G/N with $H/N \trianglelefteq G/N$. Since H is normal in G, $G = [H]K$ for some subgroup K of G. By Lemma 1.11, it follows that $G/N = [H/N]KN/N$ or that G/N splits over H/N.

The proof of (iv) is almost identical to the proof of (ii) Lemma 8.21 and is omitted. □

Given Theorem 8.26, a stronger version of (ii) in Lemma 8.27 can be stated for nC-groups.

Theorem 8.28. *Every normal subgroup of a nC-group is an nC-group.*

Now, to examine the structure of nC-groups.

Theorem 8.29. [Satz 8 in Gaschütz [41]] *A group G with elementary abelian Sylow subgroups splits over every normal subgroup.*

Beweis. Let N be a normal subgroup of G. If $N = \{1\}$ or G, then G clearly splits over N. Suppose N is a proper nontrivial normal subgroup of G. First, consider the case that N is nilpotent. Since N has abelian Sylow subgroups, N must itself be abelian. Let p be a prime such that p divides the order of N, and let P be a Sylow p-subgroup of G.

Given that P is elementary abelian, P trivially splits over $P \cap N$. Thus, by Theorem 4.4, G splits over N.

Now, suppose that N is not nilpotent and proceed by induction on the order of G. Since N is not nilpotent, there exists a Sylow p-subgroup P of N, for some prime p, such that $\mathcal{N}_G(P) \neq G$. By the Frattini argument (Lemma 1.7), $G = N\mathcal{N}_G(P)$. Given that $N \cap \mathcal{N}_G(P) \trianglelefteq \mathcal{N}_G(P)$, it follows by induction that $\mathcal{N}_G(P) = [N \cap \mathcal{N}_G(P)]H$ for $H \leq \mathcal{N}_G(P)$. Consequently,

$$G = N\mathcal{N}_G(P) = N((N \cap \mathcal{N}_G(P))H) = NH$$

with $N \cap H \leq N \cap \mathcal{N}_G(P) \cap H = \{1\}$. This means G splits over N. □

It should be noted that the converse of this theorem is not true as shown by the symmetric group $\mathcal{S}_4$.

Theorem 8.30. [Satz 6.6 in Gaschütz [43]] *The trivial subgroup* $\{1\}$ *is a prefrattini subgroup of G if and only if G is an nC-group.*

Beweis. (⇒). Let N be a normal subgroup of G. Since $\{1\}$ is a prefrattini subgroup of G, Theorem 1.66 implies all chief factors of G contained in N are complemented. Thus, by Lemma 5.2, N has a complement in G.

(⇐) Let

$$\{1\} = N_0 \lhd N_1 \lhd \cdots \lhd N_{r-1} \lhd N_r = G$$

be a chief series for G, and let W be a prefrattini subgroup of G. Since N_{r-1} is complemented in G, G/N_{r-1} is a complemented chief factor (use Lemma 1.11). Thus, by Theorem 1.66, W avoids G/N_{r-1}. This means $W \cap G \leq N_{r-1}$. Proceed in this manner to obtain $W = \{1\}$. □

Bechtell [8] and Christensen [28, 29] completed lengthy investigations of nC-groups. Both mainly studied solvable nC-groups and many of their results are presented here. However, before getting to them, a few more general results are given.

Lemma 8.31. [Lemma 4.1 in Christensen [28]] *Every nC-group is a splitting extension of a perfect cC-group and a solvable nC-group.*

Beweis. Let G be an nC-group and consider the derived series for G. For some $n \geq 1$, $G^n = G^{n+i}$ for $i \geq 0$. Thus, G^n is perfect, and by Lemma 8.27 is a cC-group. In addition, G splits over G^n as G is an nC-group. Furthermore, by Lemma 8.27, G/G^n is a solvable nC-group. □

Lemma 8.32. [Special Case of Lemma 4.2 in Christensen [28]] *Every nilpotent nC-group is elementary abelian.*

Beweis. Let P be a Sylow p-subgroup, for some prime p, of a nilpotent nC-group G. Since P is normal in G, Corollary 1.54 implies $\Phi(P) \leq \Phi(G)$, where $\Phi(G) = \{1\}$ (by Lemma 8.27). Thus, $\Phi(P) = \{1\}$ and G is elementary abelian. □

As mentioned above, much more attention has been paid to solvable nC-groups than to general nC-groups. The turn to solvable nC-groups begins with the following result.

Theorem 8.33. [Gross [47]] *A solvable group is an nC-group if and only if it is a $\mathcal{K}$-group.*

Beweis. ($\Rightarrow$)Let G be a solvable nC-group and proceed by induction on the order of G. Since G is solvable, $\mathcal{F}(G) \neq \{1\}$, and since G is an nC-group, $\Phi(G) = \{1\}$. Thus, by Theorem 1.69, $\mathcal{F}(G)$ is the direct product of minimal normal subgroups of G. By Lemma 1.25 and Lemma 8.27, $G/\mathcal{F}(G)$ is a solvable nC-group. By induction, $G/\mathcal{F}(G)$ is a $\mathcal{K}$-group. However, since G is an nC-group, $G = [\mathcal{F}(G)]K$ for some subgroup K of G with $K \cong G/\mathcal{F}(G)$. Thus, by Theorem 8.3, G is a $\mathcal{K}$-group.

($\Leftarrow$) The converse is trivial. □

Using Theorem 8.9, the following corollary follows directly from Theorem 8.33.

Corollary 8.34. [Gross [47]] *Let G be a supersolvable group. Then, G is an nC-group if and only if G is an aC-group*

Theorem 8.35. [Theorem 1.3 in Bechtell [8]] *The class of solvable nC-groups is a formation.*

Beweis. By Lemma 1.25 and Lemma 8.27, G/N is a solvable nC-group for each normal subgroup N of G. The second criteria follows from Theorem 8.8 and Theorem 8.33. □

It should be noted that Juan-Martínez proved in [68] that the word solvable can be removed from Theorem 8.35. In other words, for an arbitrary group G

(i) if G is an nC-group with $N \trianglelefteq G$, then G/N is an nC-group, and
(ii) if G/N and G/M are nC-groups for $N, M \trianglelefteq G$, then $G/(N \cap M)$ is an nC-group.

Let $\mathfrak{nC}$ denote the formation of solvable nC-groups. Now, let $\mathfrak{F}$ be an arbitrary formation. Then, $c(\mathfrak{F})$ is the collection of all groups for which the $\mathfrak{F}$-residual $G_{\mathfrak{F}}$ of a group G has only complementable chief factors of G contained in it. Blessenohl and Brewster [17] proved that $c(\mathfrak{F})$ is a formation.

Theorem 8.36. [Theorem 5 in Bechtell [13]] *Let $\mathfrak{F}$ be a formation, and let $G \in c(\mathfrak{F})$ be solvable. Then, $G_{\mathfrak{F}}$ is a solvable nC-group and G splits over $G_{\mathfrak{F}}$.*

Bechtell's Proof. The fact that G splits over $G_{\mathfrak{F}}$ follows from Lemma 5.2. Suppose $(G_{\mathfrak{F}})_{\mathfrak{nC}} \neq \{1\}$ and proceed by induction on the order of G. Let N be a minimal normal subgroup of G, such that $N \leq (G_{\mathfrak{F}})_{n\mathfrak{C}}$. Given that $(G/N)_{\mathfrak{F}} = G_{\mathfrak{F}}/N$ by Lemma 1.93, it follows by induction that $G_{\mathfrak{F}}/N \in n\mathfrak{C}$. Consequently $(G_{\mathfrak{F}})_{n\mathfrak{C}} = N$ and $\Phi(G_{\mathfrak{F}}) \leq N$.

By Theorem 1.66, $W \leq N$ for a prefrattini subgroup W of $G_{\mathfrak{F}}$. Suppose $\Phi(G_{\mathfrak{F}}) \neq \{1\}$. Then, by Theorem 1.67 and Corollary 1.54, $\Phi(G_{\mathfrak{F}}) \leq N \cap \Phi(G) \neq \{1\}$. This means $G_{\mathfrak{F}}$ contains a Frattini chief factor of G, a contradiction. Thus, $\Phi(G_{\mathfrak{F}}) = \{1\}$. Given that N is abelian and $N \cap \Phi(G_{\mathfrak{F}}) = \{1\}$, $G_{\mathfrak{F}}$ splits over N by Theorem 3.9. By Theorem 1.66, W avoids N. This means $W = \{1\}$ and that $(G_{\mathfrak{F}})_{n\mathfrak{C}} = \{1\}$ by Theorem 8.30. □

As mention before, Christensen [28, 29], in a series of two papers, completed an in-depth investigation of solvable nC-groups. A number of the results he obtained are listed here.

Lemma 8.37. [Lemma 4.3 in Christensen [28]] *For all x, if G is an xC-group which is solvable of length n, G may be expressed in the form*

$$G = A_1 A_2 \cdots A_n,$$

where each A_i, for $1 \leq i \leq n$, is an abelian cD-group and satisfies the conditions
(i) $A_i \cap A_{i+1} \cdots A_n = \{1\}$, *for* $i = 1, \ldots, n-1$, *and*
(ii) $A_{i+r} \leq \mathcal{N}_G(A_i)$, *for* $0 < i < n$ *and* $0 < r < n - i$.

In particular, we may choose $G^{n-i} = A_1 \cdots A_i$ for $0 < i \leq n$.

Christensen's Proof. The result will be proven for $x = f$ as all xC-groups are fC-groups. It is fairly simple to show that G^{n-1} is an fC-group. Given that G^{n-1} is abelian, it is an abelian cD-group. Since G is an fC-group, there is a subgroup H_1 of G, such that $G = [G^{n-1}]H_1$, where H_1 is solvable fC-group of length $n-1$.

Applying this process again, $H_1 = [H_1^{n-2}]H_2$, where H_1^{n-2} is an abelian cD-group and H_2 is a solvable fC-group of length $n - 2$. With $H_0 = G$, complete this process n-times to obtain $G = H_0^{n-1} H_1^{n-2} H_2^{n-3} \cdots H_{n-1}^0$ with each H_{i-1}^{n-i} an abelian cD-group. Letting $A_i = H_{i-1}^{n-i}$, the expansion is obtained with $A_{i+r} \leq \mathcal{N}_G(A_i)$, for $0 < r < n - i$, $0 < i < n$, and $A_i \cap A_{i+1} \cdots A_n = \{1\}$ for $1 < i < n$. It also follows that $G^{n-i} = A_1 \cdots A_i$ for $1 < i \leq n$. □

Lemma 8.37 motivates the following definition.

Definition 8.38. A **complementing expansion** of a group G is defined to be an expression of G in the form $G = A_1 A_2 \cdots A_n$, where each A_i is an abelian cD-group such that $A_{i+r} \leq \mathcal{N}_G(A_i)$ for $n - i \geq r \geq 0$ and $A_i \cap A_{i+1} \cdots A_n = \{1\}$ for $i = 1, \ldots, n-1$.

All groups with a complementing expansions of length n are solvable of at most length n with the possibility that $G^{n-i} = A_1 \cdots A_i$. This motivates the next definition.

Definition 8.39. A **derived expansion** of a solvable group G is a complementing expansion of G, such that $G = A_1 \cdots A_n$, where $G^{n-i} = A_1 \cdots A_i$, with n the solvability length of G, for $i = 1, 2, \ldots, n$.

Lemma 8.40. [Lemma 5.1 in Christensen [28]] *If $G = A_1 \cdots A_n$ is a complementing expansion of a solvable group G, then*
(i) $A_{i+r} \le \mathcal{N}_G(A_{ij})$ *for $n - i \ge r \ge 0$, with $j = 1, \ldots, t$, where A_{ij} is the Sylow p_j-subgroup of A_i and $p_1, \ldots, p_t$ are all divisors of $|G|$;*
(ii) *the product $S_j = A_{1j} \cdots A_{nj}$ (where the actual order of the A_{ij} in the product is immaterial) is a Sylow p_j-subgroup of G;*
(iii) *the set $\{S_1, \ldots, S_t\}$ forms a Sylow basis for G.*

Christensen's Proof. For (i), first note that A_i abelian implies A_{ij} is characteristics in A_i. Furthermore, by the definition of a complementing expansion (Definition 8.38), $A_{i+r} \le \mathcal{N}_G(A_i)$ for $0 \le r \le n - i$. This implies $A_{i+r} \le \mathcal{N}_G(A_{ij})$ for $0 \le r \le n - i$ and $j = 1, 2, \ldots, t$.

To prove (ii), apply condition (i) to get $A_{(i+r)k} \le \mathcal{N}_G(A_{ij})$ for all i, j, k, and r. This implies (using Lemma 1.4) that

$$A_{ux}A_{vy} = A_{vy}A_{ux}$$

for $u, v = 1, 2, \ldots, n$ and $x, y = 1, 2, \ldots, t$. Given that

$$G = A_1 \cdots A_n = \prod_{i=1}^{n} (A_{i1} \cdots A_{it})$$

the previous observation results in

$$G = \prod_{j=1}^{t} (A_{1j} \cdots A_{nj}) = \prod_{j=1}^{t} \left(\prod_{i=1}^{n} A_{ij} \right),$$

and the fact that $\prod_{i=1}^{n} A_{ij}$ is a subgroup of G.

Since $|G| = \prod_{i=1}^{n} |A_i|$ with $|A_i| = \prod_{j=1}^{t} |A_{ij}|$, it follows that

$$|G| = \prod_{i=1}^{n} \left(\prod_{j=1}^{t} |A_{ij}| \right) = \prod_{j=1}^{t} \left(\prod_{i=1}^{n} |A_{ij}| \right) = \prod_{j=1}^{t} p_j^{r_j},$$

where $p_j^{r_j} = \prod_{i=1}^{n} |A_{ij}|$. This means the Sylow p_j-subgroups of G are of order $p_j^{r_j}$ and that $S_j = \prod_{i=1}^{n} A_{ij}$ is a Sylow p_j-subgroup of G.

For (iii), the fact that the Sylow subgroups S_j pairwise permute results in the set $\{S_1, \ldots, S_t\}$ forming a Sylow basis for G. □

Now, a necessary and sufficient condition can be given for a solvable group to be an nC-group.

Theorem 8.41. [Theorem 5.2 in Christensen [28]] *A necessary and sufficient condition for a solvable group G to be an nC-group is that it should possess at least one complementary expansion $G = A_1 \cdots A_n$ and that all such expansions should satisfy the following two conditions:*
(i) *If $N \lhd G$ and A_{ij} is the Sylow p_j-subgroup of A_i for $i = 1, \dots, n$, then $(N \bigcap \prod_{i=1}^{n} A_{ij})$ is complemented in $\prod_{i=1}^{n} A_{ij}$ for all j.*
(ii) *The complements in (i) may be chosen in such a way that their product (in any order) is a subgroup.*

Christensen's Proof. ($\Rightarrow$) Let G be a solvable nC-group of solvability length n. By Lemma 8.37, G has a complementing expansion $G = A_1 \cdots A_n$, where each A_i, for $1 \le i \le n$ is an abelian cD-group such that $A_{i+r} \le \mathcal{N}_G(A_i)$, for $0 \le r \le n-i$, and $A_i \cap A_{i+1} \cdots A_n = \{1\}$ for $i = 1, 2, \dots, n-1$. Let N be a normal subgroup of G. By Lemma 8.40, $S_j = \prod_{i=1}^{n} A_{ij}$, where A_{ij} is the Sylow p_j-subgroup of A_i, is a Sylow p_j-subgroup of G. By Theorem 4.12, $N \bigcap \prod_{i=1}^{n} A_{ij}$ is complemented in $\prod_{i=1}^{n} A_{ij}$. It remains to show that condition (ii) holds.

First, note that N will be solvable of solvability length at most n. Let $B_1 = N^{n-1}$. This implies $B_1 \lhd G$ and that $G = [B_1]K_1$ for some subgroup K_1 of G. Thus, $K_1 \le \mathcal{N}_G(B_1)$ and $B_1 \le N \le B_1K_1$, which results in (using the modular identity (Lemma 1.6)) $B_1(N \cap K_1) = B_1K_1 \cap N = N$. Note that $N \cap K_1$ is solvable of solvability length at most $n-1$. Given that $N \cap K_1 \lhd K_1$, then $B_2 = (N \cap K_1)^{n-2}$ is normal in K_1. By Lemma 8.27, K_1 is an nC-group. Thus, $K_1 = [B_2]K_2$ for $K_2 \le K_1$ and $K_2 \le \mathcal{N}_G(B_2)$. This implies $G = B_1K_1 = B_1B_2K_2$ with $B_1B_2(N \cap K_2) = B_1B_2K \cap N = N$, where $N \cap K_2$ is solvable of solvability length at most $n-2$. Again, using Lemma 8.27, it follows that K_2 is an nC-group.

Repeat this process n-times to obtains $G = B_1B_2 \cdots B_nK_n$, where $B_i \cap B_{i+1} \cdots B_nK_n = \{1\}$, for $i = 1, \dots, n-1$, $B_{i+r} \le \mathcal{N}_G(B_i)$, for $i = 1, 2, \dots, n$ and $0 \le r \le n-i$, and $K_n \le \mathcal{N}_G(B_i)$ for $i = 1, 2, \dots, n$. Given that $B_1 \cdots B_i = N^{n-i}$, $N = B_1 \cdots B_n$ is a derived expansion for N.

Given that K_n is a solvable nC-group, it follows from Lemma 8.37 that $K_n = C_1 \cdots C_n$ has a derived expansion. Given that $K_n \le \mathcal{N}_G(B_i)$, for $1 \le i \le n$, $C_k \le \mathcal{N}_G(B_1)$ for $1 \le k \le n$. In addition, it follows that $B_i \cap B_{i+1} \cdots B_nC_1 \cdots C_n = \{1\}$ for $1 \le i \le n-1$. Since each B_i and C_k, for $1 \le i, k \le n$ are cD-groups, $G = B_1 \cdots B_nC_1 \cdots C_n$ is a complementing expansion of G.

Let B_{ij} and C_{ij}, for $1 \le i \le n$ be the unique Sylow p_j-subgroups of B_i and C_i, for $1 \le j \le t$, respectively. By Lemma 8.40, the collection of $S_j = \prod_{i=1}^{n} B_{ij} \prod_{i=1}^{n} C_{ij}$, for $j = 1, \dots, t$, is a Sylow basis for G, with $\prod_{i=1}^{n} B_{ij}$ and $\prod_{i=1}^{n} C_{ij}$ being Sylow p_j-subgroups of N and K respectively. This results in

$$(N \cap S_j) = \prod_{i=1}^{n} B_{ij} \text{ and } (K \cap S_j) = \prod_{i=1}^{n} C_{ij}$$

or that

$$S_j = (N \cap S_j)(K_n \cap S_j)$$

for $1 \le j \le t$.

By Lemma 8.40, if T_j is the product of Sylow p_j-subgroups of the components of a complementing expansion of G, then $\{T_1, \ldots, T_t\}$ is a Sylow basis for G. Given that any two Sylow bases are conjugate (see [34]), there exists an element $g \in G$, such that $S_j^g = T_j$ for $1 \le j \le t$. This implies

$$T_j = (N \cap T_j)(K_n^g \cap T_j)$$

with

$$N = \prod_{j=1}^{t} (N \cap T_j) \text{ and } \prod_{j=1}^{t} (K_n^g \cap T_j) = K_n^g.$$

Given that K_n^g is a subgroup of G, item (ii) is now proven.

($\Leftarrow$) Assume that $G = A_1 \cdots A_n$ has a complementing expansion that satisfies conditions (i) and (ii). Let $T_j = \prod_{i=1}^{n} A_{ij}$ be the product of the Sylow p_j-subgroups of A_i. By Lemma 8.40, $\{T_1, \ldots, T_t\}$ is a Sylow basis for G with $G = \prod_{j=1}^{t} T_j$. Now, let N be a normal subgroup of G. By conditions (i) and (ii), $T_j = [N \cap T_j]K_j$ for some subgroup K_j of G, where $\prod_{j=1}^{t} T_j$ is a subgroup of G. This means $G = \prod_{j=1}^{t} (N \cap T_j) K_j$.

Consider the subgroups $\prod_{j=1}^{t} (N \cap T_j)$ and $\prod_{j=1}^{t} K_j$ of G. If there is a nontrivial element $g \in \prod_{j=1}^{t} (N \cap T_j) \bigcap \prod_{j=1}^{t} K_j$, then $g = k_1 \cdots k_t$, where $k_j \in K_j$ for $1 \le j \le t$. However, given that $g \in \prod_{j=1}^{t} (N \cap T_j)$, by Lemma 1.37, it follows that $k_j \in N \cap T_j$. However, given that $N \cap T_j \cap K_j = \{1\}$, this implies $k_j = 1$ for $1 \le j \le t$. This results in $\prod_{j=1}^{t} (N \cap T_j) \bigcap \prod_{j=1}^{t} K_j = \{1\}$. This fact shows

$$\begin{aligned}
\left|\prod_{j=1}^{t} (N \cap T_j) \prod_{j=1}^{t} K_j\right| &= \left|\prod_{j=1}^{t} (N \cap T_j)\right| \left|\prod_{j=1}^{t} K_j\right| \\
&= \prod_{j=1}^{t} |N \cap T_j| \prod_{j=1}^{t} |K_j| \\
&= \prod_{j=1}^{t} |N \cap T_j|\,|K_j| \\
&= \prod_{j=1}^{t} |(N \cap T_j) K_j| \\
&= \prod_{j=1}^{t} |T_j| \\
&= |G|.
\end{aligned}$$

As a result, $G = \left[\prod_{j=1}^{t}\left(N \cap T_j\right)\right]\prod_{j=1}^{t} K_j$. □

Christensen then used Theorem 8.41 to prove a number of results concerning nC-groups, one of which is listed here.

Theorem 8.42. [Theorem 5.4(i) in Christensen [28]] *A metabelian group is an nC-group if and only if it is the splitting extension of two abelian groups and has elementary abelian Sylow subgroups.*

In 1967, Christensen continued the work he started in [28]. He proved many interesting results, two of which are included here. The first is a parallel result to Theorem 8.29 by Gaschütz.

Theorem 8.43. [Theorem 4.2 in Christensen [29]] *Every group with abelian Sylow subgroups and a complementing expansion is an nC-group.*

Theorem 8.44. [Theorem 5.5 in Christensen [29]] *Let G be a nC-group of solubility length n. Then, all complements of G^{n-1} are conjugate.*

This section ends with a brief examination of a class of groups very similar to nC-groups. the motivation for this class comes from the fact that if N is a normal subgroup of a group G with $N \leq \Phi(G)$, then N will not have a complement in G.

Definition 8.45. A group G is a $\mathcal{B}$-group if G splits over every normal subgroup where $N \nleq \Phi(G)$.

This class of groups was introduced by Bechtell in [6] where he studied supersolvable $\mathcal{B}$-groups. While not going into great detail, a few properties of $\mathcal{B}$-groups will be established.

Lemma 8.46. *Let G be a $\mathcal{B}$-group.*
(i) *Any homomorphic image of G is a $\mathcal{B}$-group.*
(ii) $\Phi(G)N/N = \Phi(G/N)$.
(iii) *If $N \trianglelefteq G$, then either $N \leq \Phi(G)$ or $\Phi(G) < N$.*

Beweis. For (i), let N be normal in G, and let $M/N \trianglelefteq G/N$, such that $M/N \nleq \Phi(G/N)$. Given that $\Phi(G)N/N \leq \Phi(G/N)$ by Lemma 1.52, it follows that $M \nleq \Phi(G)$. This means $G = [M]H$ for $H \leq G$. Thus, by Lemma 1.11, $G/N = [M/N]HN/N$.

To prove (ii), first note that by Lemma 1.52, $\Phi(G)N/N \leq \Phi(G/N)$. Suppose that $\Phi(G)N/N < \Phi(G/N)$. Letting $M/N = \Phi(G/N)$, it follows that $M \triangleleft G$ with $M \nleq \Phi(G)$. This means $G = [M]H$ for $H \leq G$. Thus, by Lemma 1.11, $G/N = [M/N]HN/N$, which is a contradiction.

For (iii), proceed by assuming that $N \nleq \Phi(G)$ and that $\Phi(G) \nless N$. It follows that $G = [N]H$ for $H \leq G$. If $\Phi(H) \neq \{1\}$, then, since $N\Phi(H)$ is normal in G, $G = [N\Phi(H)]K$ for $K \leq G$ and $|K| < |H|$.

By Lemma 1.11, it follows that $G/N = [N\Phi(H)/N]KN/N$. Given that $G/N \cong H$, item (i) implies that H is a $\mathcal{B}$-group. This means, using item (ii), that $N\Phi(H)/N = \Phi(HN/N)$ and that $KN/N \cong H$. However, since $KN/N \cong K/K \cap N = K$, this means that $|K| = |H|$, a contradiction. Therefore, $\Phi(H) = \{1\}$, which, since $\Phi(G)N/N = \Phi(G/N)$, is another contradiction. Thus, either $N \leq \Phi(G)$ or $\Phi(G) < N$. □

Lemma 8.47. *Let G be a supersolvable $\mathcal{B}$-group with $\Phi(G) \neq \{1\}$. Then, the Fitting subgroup $\mathcal{F}(G)$ is the Sylow p-subgroup for p the largest prime dividing the order of G.*

Beweis. By Theorem 1.43, the Sylow p-subgroup P, for p the largest prime diving the order of G, is normal in G. This means $P \leq \mathcal{F}(G)$. Suppose that $P < \mathcal{F}(G)$. Since $\mathcal{F}(G)$ is nilpotent, there exists a q-group Q, for q a prime not equal to p, such that $Q \lhd G$.

By Lemma 8.46, either $P \leq \Phi(G)$ or $\Phi(G) < P$. Since $P \nleq \Phi(G)$ (see Lemma 1.51), it follows that $\Phi(G) < P$ or that $\Phi(G)$ is a p-group. However, by Lemma 8.46, it also follows that either $Q \leq \Phi(G)$ or $\Phi(G) < Q$. This contradiction implies that $P = \mathcal{F}(G)$. □

Theorem 8.48. [Proposition 1.4 in Bechtell [6]] *Let G be a supersolvable group. For $H \neq \{1\}$ and $K \neq \{1\}$, $G = H \times K$ is a $\mathcal{B}$-group if and only if H and K are aC-groups.*

Beweis. ($\Rightarrow$) By Lemma 8.46, $H \leq \Phi(G)$ or $\Phi(G) < H$. The decomposition of G forces $\Phi(G) < H$. A similar argument results in $\Phi(G) < K$. Given that $H \cap K = \{1\}$, it follows that $\Phi(G) = \{1\}$. Thus, by Corollary 1.54, $\Phi(H) = \Phi(K) = \{1\}$. Furthermore, by Lemma 8.46, H and K are supersolvable $\mathcal{B}$-groups. Therefore, by Corollary 8.34, H and K are aC-groups.

($\Leftarrow$) Suppose that H and K are aC-groups. Then, by Lemma 8.21, G is an aC-group. This implies that G is a $\mathcal{B}$-group. □

Now, let G be a supersolvable $\mathcal{B}$-group and consider $Z(G)$. By Lemma 8.46, $Z(G) \leq \Phi(G)$ or $\Phi(G) < Z(G)$. If $\Phi(G) < Z(G)$, then $G = Z(G) \times H$ for $H \leq G$. It follows that $Z(H) = \{1\}$ and by Theorem 8.48 that H is an aC-group. Given that $\Phi(Z(G)) = \{1\}$ (again use Theorem 8.48), $Z(G)$ is the direct product of elementary abelian p-groups. Next, consider a nilpotent $\mathcal{B}$-group G. If the order of G is divisible by two or more primes, then Theorem 8.48 implies that G is the direct product of elementary abelian p-groups. This motivates the following lemma and theorem.

Lemma 8.49. *Let G be a supersolvable nonnilpotent $\mathcal{B}$-group such that $\Phi(G) \neq \{1\}$ and G is not the direct product of two nontrivial proper subgroups.*

(i) $\Phi(G) < G' \leq \mathcal{F}(G)$.
(ii) *If* $|\Phi(G)| = p$, *for* p *a prime, then* $\Phi(G)$ *is the unique minimal normal subgroup of* G.
(iii) $\mathcal{F}(G)$ *is the Sylow* p*-subgroup for the largest prime* p *dividing the order of* G *with* $p \neq 2$.

Beweis. By Theorem 1.44, G' is nilpotent and $G' \leq \mathcal{F}(G)$. By Lemma 8.46, $G' \leq \Phi(G)$ or $\Phi(G) < G'$. If $G' \leq \Phi(G)$, then G is nilpotent by Theorem 1.58. This contradiction implies $\Phi(G) < G'$, which completes the proof of (i).

For (ii), let N be a minimal normal subgroup of G. By Lemma 8.46, $N \leq \Phi(G)$ or $\Phi(G) < N$. Since $|\Phi(G)| = p$, it follows that $N = \Phi(G)$ and that $\Phi(G)$ is the unique minimal normal subgroup of G.

To prove (iii), first note that by Lemma 8.47, $\mathcal{F}(G)$ is the Sylow p-subgroup of G for p the largest prime dividing the order of G. If $p = 2$, then G is nilpotent, which is a contradiction. This implies $p \neq 2$. □

The examination of $\mathcal{B}$-groups done so far motivates the following definition.

Definition 8.50. A group G is a ***B_1*-group** if it has the following properties.
(i) G is a nonnilpotent supersolvable group that can't be expressed as the product of two of its nontrivial proper subgroups;
(ii) $\mathcal{F}(G)$ is the Sylow p-subgroup of G, where p is the largest prime dividing the order of G with $p \neq 2$;
(iii) $\Phi(G) < G' \leq \mathcal{F}(G)$; and
(iv) $\Phi(G)$ is the unique minimal normal subgroup of G.

Theorem 8.51. [Theorem 2.1 in Bechtell [6]] *A* B_1*-group* G *is a* $\mathcal{B}$*-group if and only if* G *satisfies one of the following conditions:*
(i) $G = [\mathcal{F}(G)]C$, *for* $\mathcal{F}(G)$ *cyclic of order* p^2, C *cyclic of square-free order acting faithfully on* $\mathcal{F}(G)$, *and* $p \nmid |C|$.
(ii) $G = [\mathcal{F}(G)]C$, *where* $\mathcal{F}(G)$ *is the nonabelian* p*-group of order* p^3 *and exponent* p, *and* C *is the direct product of two cyclic groups of square-free order having at least one factor different from the identity, such that* C *acts faithfully on* $\mathcal{F}(G)$, C *fixes at least one maximal subgroup of* $\mathcal{F}(G)$ *(and hence at least two), and* $p \nmid |C|$.

Bechtell's Proof. Since $\Phi(\mathcal{F}(G)) \leq \Phi(G)$ by Corollary 1.54 with $\Phi(G)$ the unique minimal normal subgroup of G, $\Phi(\mathcal{F}(G)) = \Phi(G)$. Given that $\mathcal{F}(G)$ is the Sylow p-subgroup of G for p the largest prime dividing the order of G, with $|\Phi(G)| = p$ and $\Phi(G) < G'$, it follows that there exists a normal subgroup N of G, such that $\Phi(G) < N \leq G'$ with $|N| = p^2$. Since G is a $\mathcal{B}$-group, $G = [N]H$ for $H \leq G$. However, since H is a supersolvable $\mathcal{B}$-group with $\Phi(H) = \{1\}$ (by Lemma 8.46), H is an aC-group by Corollary 8.34. By Lemma 1.11, $\mathcal{F}(G) = [N](\mathcal{F}(G) \cap H)$, with $\mathcal{F}(G) \cap H$ an elementary p-group.

Case 1: $(\mathcal{F}(G))' = \{1\}$.

In this case, $\mathcal{F}(G) = N \times (\mathcal{F}(G) \cap H)$. Given that $\Phi(\mathcal{F}(G)) \neq \{1\}$, this implies that N is cyclic of order p^2. If $\mathcal{F}(G) \cap H \neq \{1\}$, the elements of order p in $\mathcal{F}(G)$ form a characteristic subgroup F of $\mathcal{F}(G)$, such that $[\mathcal{F}(G) : F] = p$. Given that $F \nleq \Phi(G)$, $G = [F]K$ for $K \leq G$. However, since K is a supersolvable $\mathcal{B}$-group with $\Phi(K) = \{1\}$, K is an aC-group by Corollary 8.34. By Lemma 1.11, it follows that $\mathcal{F}(G) = [F](\mathcal{F}(G) \cap K)$ and that $\mathcal{F}(G)$ is elementary abelian. This contradiction implies $\mathcal{F}(G) \cap H = \{1\}$ and $\mathcal{F}(G) = N$, with $\mathcal{F}(G)$ cyclic of order p^2. This results in $G = [\mathcal{F}(G)]H$ with H an aC-group. Since $Aut(\mathcal{F}(G))$ is cyclic, H is cyclic. By Theorem 8.22, H is of square-free order with $p \nmid |H|$ and it acts faithfully on $\mathcal{F}(G)$.

For the converse, suppose that $G = [\mathcal{F}(G)]C$, with $\mathcal{F}(G)$ cyclic of order p^2, for p the largest prime dividing the order of G, C is cyclic of square-free order acting faithfully on $\mathcal{F}(G)$, and $p \nmid |C|$. Let M be normal in G, such that $M \nleq \Phi(G)$. Since $\Phi(G) < M$, consider the group $G/\Phi(G)$. Given that $G/\Phi(G)$ is a group of square-free order, $G/\Phi(G)$ is an aC-group by Lemma 8.21. This implies $G/\Phi(G) = [M/\Phi(G)]L/\Phi(G)$ for $L/\Phi(G) \leq G/\Phi(G)$. However, L is also a group of square-free order so $L = \Phi(G)R$ for $R < L$. Consequently, $G = [M]R$ and G splits over M.

Case 2: $(\mathcal{F}(G))' = \Phi(\mathcal{F}(G)) = \Phi(G) \neq \{1\}$.

If $(\mathcal{F}(G))' < Z(\mathcal{F}(G))$, then $\mathcal{F}(G)$ is abelian, a contradiction. This means $(\mathcal{F}(G))' = Z(\mathcal{F}(G)) = \Phi(\mathcal{F}(G))$. Furthermore, since $ker(\alpha) \leq Z(\mathcal{F}(G)) \leq N$, for the homomorphism $\alpha : \mathcal{F}(G) \cap H \to Aut(G)$, it follows that $|\mathcal{F}(G) \cap H| = p$ and that $|\mathcal{F}(G)| = p^3$.

Suppose $\mathcal{F}(G)$ does not have exponent p. Then, there exists a subgroup S of $\mathcal{F}(G)$ generated by the elements of order p in $\mathcal{F}(G)$, such that $[\mathcal{F}(G) : S] = p$. Since S is characteristics in $\mathcal{F}(G)$, it is normal in G. This means $G = [S]L$ for $L \leq G$. However, by Lemma 1.11, this means $\mathcal{F}(G) = [S](\mathcal{F}(G) \cap L)$, which is a contradiction.

Now, suppose that $\mathcal{F}(G)$ has exponent p. By Lemma 1.11, $\mathcal{F}(G)$ will split over any of its subgroups, except for $\Phi(\mathcal{F}(G))$, that is also normal in G. By Theorem 1.69, $\mathcal{F}(G/\Phi(G)) = \mathcal{F}(G)/\Phi(G)$, which implies $\mathcal{F}(G)/\Phi(G)$ is elementary abelian of order p^2 and that $C_{G/\Phi(G)}(\mathcal{F}(G)/\Phi(G)) = \mathcal{F}(G)/\Phi(G)$. Furthermore, given that $G = [\mathcal{F}(G)]C$, Lemma 1.11 implies $G/\Phi(G) = [\mathcal{F}(G)/\Phi(G)]C\Phi(G)/\Phi(G)$, where $C\Phi(G)/\Phi(G) \cong C$. Using Maschke's theorem (Theorem 4.4.1 in [7]), $\mathcal{F}(G)/\Phi(G)$ is completely reducible with respect to $C\Phi(G)/\Phi(G)$ with $C\Phi(G)/\Phi(G)$ acting faithfully on $\mathcal{F}(G)/\Phi(G)$.

Now, note that $\mathcal{F}(G)/\Phi(G) = F_1/\Phi(G) \times F_2/\Phi(G)$, where $F_1/\Phi(G)$ and $F_2/\Phi(G)$ are normal in $G/\Phi(G)$. Since $\mathcal{F}(G) \cap H$ is the kernel of the homomorphism $\beta : C\Phi(G)/\Phi(G) \to Aut(F_1/\Phi(G))$ with $\mathcal{F}(G) \cap H \neq \{1\}$, it follows that no nonidentity element in $\mathcal{F}(G) \cap H$ can centralize $F_2/\Phi(G)$. Given that C is an aC-group, $C\Phi(G)/\Phi(G) = (\mathcal{F}(G) \cap H)\Phi(G)/\Phi(G) \times B/\Phi(G)$, where $B/\Phi(G) \neq \{1_{C\Phi(G)/\Phi(G)}\}$ and $(\mathcal{F}(G) \cap H)\Phi(G)/\Phi(G)$ and $B/\Phi(G)$ are cyclic of square-free order. Furthermore, $(\mathcal{F}(G) \cap H)\Phi(G)/\Phi(G)$ centralizes $F_1/\Phi(G)$ and is faithful on $F_2/\Phi(G)$ with $B/\Phi(G)$ acting faithfully on $F_1/\Phi(G)$. Thus, $C\Phi(G)/\Phi(G)$ is elementary abelian and no Sylow subgroup has order q^n with $n \geq 2$ for q a prime and $q \neq p$.

To this point, note that if $\mathcal{F}(G)$ has exponent p, then $G = [\mathcal{F}(G)]C$, where C is faithful on $\mathcal{F}(G)$, C is the direct product of two cyclic subgroups A and B of square-free order, C fixes nontrivially two maximal subgroups F_1 and F_2 of $\mathcal{F}(G)$, such that A is faithful on F_2 and B is faithful on F_1, and $p \nmid |C|$ with p odd.

Conversely, suppose a group G satisfies the conditions just mentioned. Let M be a nontrivial proper normal subgroup of G. If $M \cap \mathcal{F}(G) = \{1\}$, then $M \leq C$ and (as C is abelian) $M \leq Z(G)$. However, this implies C does not act faithfully on $\mathcal{F}(G)$, a contradiction. Thus, $M \cap \mathcal{F}(G) \neq \{1\}$.

Case 1: $\mathcal{F}(G) \leq M$.

By an argument similar to one presented earlier in this proof, G splits over M.

Case 2: $\mathcal{F}(G) \cap M \neq M$ and $\mathcal{F}(G) \cap M \neq \mathcal{F}(G)$.

Since G is nonnilpotent, Theorem 1.47 implies that not all maximal subgroups in G are normal on G. Now, suppose M is contained in every self-normalizing maximal subgroup of G. Then, by Satz 16 in [42], M is nilpotent and $M \leq \mathcal{F}(G)$. Consequently, there exists a maximal self-normalizing subgroup T of G, such that $M \nleq T$. This implies $G = MT$. By Lemma 1.60, there exists a reduced product $G = MT_1$, such that $T_1 \leq T$ and $M \cap T_1 \leq \Phi(T_1)$. Given that $\mathcal{F}(G) \nleq T$, the Sylow p-subgroups of T are elementary abelian, which means the Sylow p-subgroups of T_1 are elementary abelian. Thus, by Theorem 8.29, T_1 splits over every normal subgroup. This implies that $\Phi(T_1) = \{1\}$ and that $G = [M]T_1$.

Case 3: $M \leq \mathcal{F}(G)$.

Given that $\Phi(G) < M$, it follows that $|M| = p^2$. If $G' = \Phi(\mathcal{F}(G))$, then $G' = \Phi(G)$ and G is nilpotent by Theorem 1.58. Thus, there exists a self-normalizing maximal subgroup R, such that $N \nleq R$ and $G = NR$. An argument similar to the one presented in Case 2 shows that G splits over M. □

More results concerning supersolvable $\mathcal{B}$-groups can be found in [6].

8.5 Splitting systems

Motivated by Christensen's complementing expansions, Bechtell [11] introduced the concept of a splitting system. Recall that a group is inseparable if it does not split over any proper nontrivial normal subgroup.

Definition 8.52. A set of inseparable subgroups $\{S_1, \dots, S_n\}$ of a group G is called a **splitting system** for G if $|S_i| \neq 1$, for $i = 1, \dots, n$, and for $j = 1, \dots, n-1$, $S_1 \cdots S_{j+1} = [S_1 \cdots S_j]S_{j+1}$. The integer n is the **length** of the splitting system.

Note that while the subgroup $S_1 \cdots S_j$ is normal in $S_1 \cdots S_{j+1}$, it need not be normal in the group G. Furthermore, if $n = 1$, then the group G is inseparable. Splitting systems and their relation to group structure have been studied by Bechtell [11] and Kirtland [71, 72].

Any separable group will have two or more splitting systems. For example, consider the group $\mathcal{S}_3 = \langle a, b \mid a^3 = b^2 = 1, ab = ba^2\rangle$. It has three splitting systems $\{\langle a\rangle, \langle b\rangle\}$, $\{\langle a\rangle, \langle ab\rangle\}$, and $\{\langle a\rangle, \langle a^2 b\rangle\}$. However, not all splitting systems for a groups need be of the same length. The dihedral group $D_8 = \langle x, y \mid x^4 = y^2 = 1, xy = yx^3\rangle$ has a splitting system $\{\langle x\rangle, \langle y\rangle\}$ of length 2 and a splitting system $\{\langle x^2\rangle, \langle y\rangle, \langle xy\rangle\}$ of length 3.

The following lemma presents a few of properties of splitting systems. It essentially follows from Lemma 1.11.

Lemma 8.53. *Let G be a group, and let N be a proper normal subgroup of G, such that $G = [N]H$ for $H < G$.*

(i) *There exists a splitting system $\{S_1, \ldots, S_n\}$ for G, such that for some i, $1 \le i \le n$, $N = S_1 \cdots S_i$ and $H = S_{i+1} \cdots S_n$.*
(ii) *If $\{T_1, \ldots, T_n\}$ is a splitting system for N, it can be extended to a splitting system $\{T_1, \ldots, T_n, T_{n+1}, \ldots, T_m\}$ for G.*
(iii) *If $\{S_1, \ldots, S_n\}$ is a splitting system for G, such that $N = S_1 \cdots S_i$ and $H = S_{i+1} \cdots S_n$, for $1 \le i \le n$, then $\{S_1, \ldots, S_i\}$ is a splitting system for N and $\{S_{i+1}, \ldots, S_n\}$ is a splitting system for H.*
(iv) *If $\{S_1, \ldots, S_n\}$ is a splitting system for N and $\{T_1, \ldots, T_m\}$ is a splitting system for H, then $\{S_1, \ldots, S_n, T_1, \ldots, T_m\}$ is a splitting system for G.*

Definition 8.54. A splitting system $\{S_1, \ldots, S_n\}$ is **complementary** if either G is inseparable or if $G = [S_1 \cdots S_j](S_{j+1} \cdots S_n)$ for $j = 1, \ldots, n-1$ and $n \ge 2$.

Lemma 8.55. *Let G be a group.*

(i) *If $\{S_1, \ldots, S_n\}$ is a complementary splitting system for G, then $\{S_{i+1}, \ldots, S_n\}$, for $1 \le i \le n$, is a complementary splitting system for $S_{i+1} \cdots S_n$.*
(ii) *If $G = [N]H$ with N inseparable and $\{S_1, \ldots, S_n\}$ a complementary splitting system for H, then $\{N, S_1, \ldots, S_n\}$ is a complementary splitting system for G.*

Beweis. For (i), first note by Lemma 8.53 that $\{S_{i+1}, \ldots, S_n\}$ is a splitting system for $S = S_{i+1} \cdots S_n$. Consider the subgroup $S_{i+1} \cdots S_j$, for $i+1 \le j \le n-1$, of S, let $s \in S_{i+1} \cdots S_j$, and let $x \in S$. Since $s \in S_{i+1} \cdots S_j$, it follows that $s = s_{i+1} \cdots s_j$, with $s_k \in S_k$ for $i+1 \le k \le j$. Since $S_1 \cdots S_j \lhd G$,

$$s^x = s_{i+1}^x \cdots s_j^x = t_1 \cdots t_j$$

with $t_l \in S_l$ for $1 \le l \le j$. This implies

$$s_{i+1}^x \cdots s_j^x t_j^{-1} \cdots t_{i+1}^{-1} = t_1 \cdots t_i.$$

Since $S_1 \cdots S_i \cap S_{i+1} \cdots S_n = \{1\}$, this is a contradiction unless $t_1 \cdots t_i = 1$. Therefore, $s^x \in S$ and $S_{i+1} \cdots S_j \lhd S$. Given that $S_{j+1} \cdots S_n < S$, it directly follows that $S = [S_{i+1} \cdots S_j](S_{j+1} \cdots S_n)$.

The proof of (ii) begins with the fact that $\{N, S_1, \ldots, S_n\}$ is a splitting system for G by Lemma 8.53. Now, consider i for $1 \le i \le n$. Given that $S_1 \cdots S_i \trianglelefteq H$, it follows that $NS_1 \cdots S_i \trianglelefteq G$. Furthermore, since $S_{i+1} \cdots S_n < G$ with $NS_1 \cdots S_i \cap S_{i+1} \cdots S_n = \{1\}$, $G = [NS_1 \cdots S_i](S_{i+1} \cdots S_n)$. This implies that $\{N, S_1, \ldots, S_n\}$ is a complementary splitting system for G. □

Bechtell, motivated by Theorem 8.9, defined complementary splitting systems and proved the following result.

Theorem 8.56. [Theorem 2 in Bechtell [11]] *A group G has a complementary splitting system consisting of groups of prime order if and only if G is a supersolvable nC-group.*

Bechtell's Proof. ($\Rightarrow$) Let $\{S_1, \ldots S_n\}$ be a complementary splitting system for G. Given that $|S_i|$ is a prime with $S_1 \cdots S_i \trianglelefteq G$ for $1 \le i \le n$, it follows that G is supersolvable.

Proceed by induction on the order of G. Clearly, the result follows in $n = 1$. Assume otherwise. Given that $G = [S_1]S_2 \cdots S_n$, Lemma 8.55 implies $\{S_2, \ldots, S_n\}$ is a complementary splitting system for $T = S_2 \cdots S_n$. By induction, T is an nC-group. By Lemma 1.52, $\Phi(G)S_1/S_1 \le \Phi(G/S_1) = \{1_{G/S_1}\}$ (G/S_1 is an nC-group). Thus, $\Phi(G) \le S_1$. Given that S_1 is of prime order with $G = [S_1]T$, it follows that $\Phi(G) = \{1\}$. Thus, by Theorem 1.69, $\mathcal{F}(G)$ is the direct product of the minimal normal subgroups of G. This means $S_1 \le \mathcal{F}(G)$ and, by the modular identity (Lemma 1.6), that

$$\mathcal{F}(G) = \mathcal{F}(G) \cap S_1 T = S_1(\mathcal{F}(G) \cap T).$$

Given that $\mathcal{F}(G) \cap T$ is normal in T, with T an nC-group, $T = [\mathcal{F}(G) \cap T]K$ for $K \le T$. Given that T is a solvable nC-group, it follows by Theorem 8.35 that K is a solvable nC-group. Given that

$$\mathcal{F}(G) \cap K \le \mathcal{F}(G) \cap K \cap T = \mathcal{F}(G) \cap T \cap K = \{1\}$$

with

$$G = S_1 T = S_1((\mathcal{F}(G) \cap T)K) = S_1(\mathcal{F}(G)K \cap T) = S_1 T \cap \mathcal{F}(G)K = \mathcal{F}(G)K,$$

it follows that $G = [\mathcal{F}(G)]K$. Thus, by Theorem 8.3 and Theorem 8.33, G is an nC-group.

($\Leftarrow$) Let G be a supersolvable nC-group, and proceed by induction on the order of G. Let p be the largest prime that divides the order of G. By Theorem 1.43, the Sylow p-subgroup S of G is normal in G. Since S is part of some chief series for G, there is a subgroup S_1 of G, such that the order of S_1 is p and $S_1 \trianglelefteq G$.

If $G = S_1$, the result holds. If not, $G = [S_1]K$ for some subgroup K of G. By Theorem 1.42 and Theorem 8.35, $G/S_1 \cong K$ is a supersolvable nC-group. Thus, by induction, K has a complementary splitting system $\{S_2, \ldots, S_n\}$, such that each subgroup in the splitting system has prime order. It directly follows from Lemma 8.55 that $\{S_1, \dots, S_n\}$ is a complementary splitting system for G with each subgroup of prime order. □

The condition of a complementary splitting system can be weakened just a bit to establish a necessary and sufficient condition for a group to be a solvable nC-group. First, a definition is needed.

Definition 8.57. A splitting system $\{S_1, \dots, S_n\}$ of a group G is **normally (characteristically) complementary** if either G is inseparable or if for each j, such that $S_1 \cdots S_j$ is normal (characteristic) in G, then $G = [S_1 \cdots S_j](S_{j+1} \cdots S_n)$.

It should be noted, as demonstrated by the group $\mathcal{S}_4$, that a group with a normally complementary splitting system or a characteristically complementary splitting system does not necessarily have a complementary splitting system.

Theorem 8.58. [Theorem 5 in Bechtell [11]] *A group G is a solvable nC-group if and only if G has a normally (characteristically) complementary splitting system that consists of groups of prime order such that all the subgroups in the splitting series which are normal (characteristic) in G form a chief (characteristic) series in G.*

Bechtell's Proof. (⇒) Let G be a solvable nC-group. Proceed by induction on the order of G. By Theorem 8.33, G is a $\mathcal{K}$-group. Thus, by Theorem 8.6, G contains a normal series

$$\{1\} = N_0 \lhd N_1 \cdots \lhd N_r = G,$$

with $r \geq 1$, such that N_{i+1}/N_i is the maximal nilpotent subgroup of G/N_i for $0 \leq i \leq r-1$. If $r = 1$, the result follows as nilpotent nC-groups are elementary abelian by Lemma 8.32.

Assume that $r \geq 2$. Then, $G = [N_1]H$ for some subgroup H of G. By Theorem 8.35, $H \cong G/N_1$ is a solvable nC-group. Thus, by induction, H has a normally (characteristically) complementary splitting system $\{S_1, \dots, S_n\}$ that consists of groups of prime order such that the subgroups in the splitting series which are normal (characteristic) in H form a chief (characteristics) series for H. Given that N_1 is also a solvable nC-group (Theorem 8.28), it follows that N_1 has a splitting system $\{T_1, \dots, T_m\}$ consists of groups of prime order such that the subgroups in the splitting series which are normal (characteristic) in N form a chief (characteristics) series for N. Given that N_1 is characteristic in G, by Lemma 8.53, the set $\{T_1, \dots, T_m, S_1, \dots, S_n\}$ is a splitting system for G consists of groups of prime order such that the subgroups in the splitting series which are normal (characteristic) in G form a chief (characteristic) series for G.

(⇐) If G is inseparable, the result trivially follows. Assume that G is separable and proceed by induction on the order of G. Let $\{T_1, \dots, T_n\}$ be a normally complementary splitting system for G, such that the subgroups in the splitting system are of prime order and that the subgroups in the splitting series which are normal in G form a chief series for G. Let i be the smallest integer, with $1 \le i \le n$, such that $H_i = T_1 \cdots T_i$ is normal in G. This means $G = [H_i](T_{i+1} \cdots T_n)$. It follows (using Lemma 8.53) that $T = T_{i+1} \cdots T_n$ has a normally complementary splitting system such that the subgroups in the splitting system are of prime order and that the subgroups in the splitting series which are normal in T form a chief series for T. By induction, T is a solvable nC-group. Given that $\Phi(G)H_i/H_i \le \Phi(G/H_i)$ by Lemma 1.52, $\Phi(G) \le H_i$. However, since H_i is minimal normal and complemented in G, this implies $\Phi(G) = \{1\}$.

Now, apply Theorem 1.69 to obtain that $\mathcal{F}(G)$ is the direct product of minimal normal subgroups of G. A simple calculation using the modular identity (Lemma 1.6) shows

$$\mathcal{F}(G) = \mathcal{F}(G) \cap G = \mathcal{F}(G) \cap H_i T = H_i(\mathcal{F}(G) \cap T).$$

Given that $\mathcal{F}(G) \cap T \lhd T$ and that T is a solvable nC-group, it follows that $T = [\mathcal{F}(G) \cap T]K$ for some subgroup K of T. Note that by Theorem 8.35 that K is also a solvable nC-group. Thus, by Lemma 1.11, $G = [\mathcal{F}(G)]K$ and is a solvable nC-group by Theorem 8.3 and Theorem 8.33.

Now, consider the case that G has a characteristically complementary splitting system $\{T_1, \dots, T_n\}$, such that it consists of groups of prime order such that all the subgroups in the splitting series which are characteristic in G form a characteristic series in G. Consider $1 \le i < j \le n$, such that $H_i = T_1 \cdots T_i$ and $H_j = T_1 \cdots T_j$ are characteristic in G with H_i the minimal nontrivial characteristic subgroup in the characteristics series given by $\{T_1, \dots, T_n\}$, and H_j/H_i is characteristically simple in G/H_i. This means that H_j/H_i is elementary p-group. Furthermore, the fact that $H_j/H_i \cap \Phi(G/H_i) = \{1_{G/H_i}\}$ implies by Theorem 1.69 that H_j/H_i is the direct product of minimal normal subgroups of G/H_i.

Now, note that $G = [H_i]T$ for $T = T_{i+1} \cdots T_n$. Given that H_i is the product of minimal normal subgroups of G, H_i has a splitting system $\{S_1, \dots, S_m\}$, such that each subgroup in it is of prime order such that subgroups in the splitting series that are normal in G form part of a chief series for G. Inductively, T has a splitting system $\{U_1, \dots, U_r\}$, such that each subgroup in it is of prime order such that subgroups in the splitting series that are normal in G form part of a chief series for G. Thus, by Lemma 8.53 $\{S_1, \dots, S_m, U_1, \dots, U_r\}$ is a normal complementary splitting system for G. Thus, by the first part of the proof, G is an nC-group.

□

Bibliography

[1] R. Baer. Nilpotent characteristic subgroups of finite groups. *Am. J. Math.*, 75:633–664, 1953.

[2] R. Baer. Classes of finite solvable groups and their properties. *Illinois J. Math.*, 1:115–187, 1957.

[3] A. Ballester-Bolinches, R. Esteban-Romero, and M. Asaad. *Products of Finite Groups. Expositions in Mathematics*. De Gruyter, Berlin, 2010.

[4] H. Bechtell. Elementary groups. *Trans. Am. Math. Soc.*, 114:355–362, 1965.

[5] H. Bechtell. A note on finite solvable $\mathcal{K}$-groups. *Proc. Am. Math. Soc.*, 17:1447–1450, 1966.

[6] H. Bechtell. A generalization of hall-complementation in finite supersolvable groups. *Trans. Am. Math. Soc.*, 140:257–270, 1969.

[7] H. Bechtell. *Theory of Groups*. Addison-Wesley, Reading, MA, 1971.

[8] H. Bechtell. On the structure of solvable nC-groups. *Rend. Sem. Mat. Univ. Padova*, 47:13–22, 1972.

[9] H. Bechtell. Inseparable finite solvable groups. *Trans. Am. Math. Soc.*, 216:47–60, 1976.

[10] H. Bechtell. Inseparable finite solvable groups ii. *Proc. Am. Math. Soc.*, 64:25–29, 1977.

[11] H. Bechtell. Splitting systems for finite solvable groups. *Arch. Math.*, 36:295–301, 1981.

[12] H. Bechtell. On nonnilpotent inseparable groups of order $p^n q^m$. *J. Algebra*, 75:223–232, 1982.

[13] H. Bechtell. Locally complemented formations. *J. Algebra*, 106:413–429, 1987.

[14] H. Bechtell. On semidirect products. *Commun. Algebra*, 19:1151–1163, 1991.

[15] R. Bercov. On groups without abelian composition factors. *J. Algebra*, 5:106–109, 1967.

[16] H. Bergström. Über abelsche erweiterungen mit zerfallenden faktorensystem. *Math. Nachr.*, 1:350–356, 1948.

[17] B. Blessenohl and B. Brewster. Über formationen und komplementierbare hauptfaktoren. *Arch. Math.*, 27:347–351, 1976.

[18] W. Burnside. *Theory of Groups of Finite Order*. Cambridge University Press, Greenwich, 1897.

[19] W. Burnside. One some properties of groups of odd order. *Proc. London Math. Soc. (1)*, 33: 162–184, 1900.

[20] W. Burnside. One some properties of groups of odd order (Second Paper). *Proc. London Math. Soc. (1)*, 33:257–268, 1900.

[21] W. Burnside. *Theory of groups of finite order*. Cambridge University Press, Cambridge, 1911.

[22] R. W. Carter. On a class of finite soluble groups. *Proc. London Math. Soc. (3)*, 9:623–640, 1959.

[23] R. W. Carter. Nilpotent self-normalizing subgroups of soluble groups. *Math. Zeitschr.*, 75: 136–139, 1961.

[24] R. W. Carter. Splitting properties of soluble groups. *J. London. Math. Soc.*, 36:89–94, 1961.

[25] R. W. Carter, B. Fischer, and T. Hawkes. Extreme classes of finite soluble groups. *J. Algebra*, 9:285–313, 1968.

[26] R. W. Carter and T. Hawkes. The $\mathfrak{F}$-normalizers of a finite solvable group. *J. Algebra*, 5:175–202, 1967.

[27] Z. Chen. Generalization of the schur-zassenhaus theorem. *J. Math. (Wuhan)*, 18:290–294, 1998.

[28] C. Christensen. Complementation in groups. *Math. Zeitschr.*, 84:52–69, 1964.

[29] C. Christensen. Groups with complemented normal subgroups. *J. London Math. Soc.*, 42: 208–216, 1967.

[30] V. Cigić. Zur $\mathfrak{F}$ komplementsatz in endlichen auflösbaren gruppen. *Radovi-Akademija Naukai Umjetnosti Bosnei Hercegovine, Odjeljenje Prirodnihi Matrmatickin Nauka*, 22:57–59, 1983.

DOI 10.1515/9783110480214-009

[31] M. Costantini and G. Zacher. The finite simple groups have complemented subgroup lattices. *Pacific J. Math.*, 213:245–251, 2004.

[32] N. Dinerstein. Finiteness conditions in groups with systems of complemented subgroups. *Math. Z.*, 106:321–326, 1968.

[33] J. D. Dixon. Complements in finite groups. *Am. Math. Month.*, 74:286–288, 1967.

[34] K. Doerk and T. Hawkes. *Finite Soluble Groups*. Walter de Gruyter, Berlin, 1992.

[35] L. Dornhoff. *m*-groups and 2-groups. *Math. Z.*, 100:226–256, 1967.

[36] W. Feit and J. Thompson. A solvability criteria for finite groups and some consequences. *Proc. Natl. Acad. Sci. USA*, 48:968–970, 1962.

[37] W. Feit and J. Thompson. Solvability of groups of add order. *Pacific J. Math.*, 13:775–1029, 1963.

[38] G. Frattini. Intorno alla generazione dei gruppi di operazioni. *Atti Reale Accad. Lincei Rend. Ser. 4*, 1:281–285, 455–456, 1885.

[39] G. Frattini. Intorno alla generazione dei gruppi d'operazioni e ad un teorema d'arimetica. *Atti Reale Accad. Lincei Rend. Ser. 4*, 2:16–19, 1886.

[40] F. G. Frobenius. Über auflösbare gruppen IV. *Sitzunsb. Kön. Preuß. Akad. Wiss. Berlin*, pages 1216–1230, 1901.

[41] W. Gaschütz. Zur erweiterungstheorie der endlichen gruppen. *J. Reine Angew. Math.*, 190:93–107, 1952.

[42] W. Gaschütz. Über die Φ-untergruppe endlicher Gruppen. *Math. Zeitschr.*, 58:160–170, 1953.

[43] W. Gaschütz. Praefrattinigruppen. *Arch. Math.*, 13:418–426, 1962.

[44] W. Gaschütz. Zur Theorie der endlichen auflösbaren Gruppen. *Math. Zeitschr.*, 80:300–305, 1963.

[45] W. Gaschütz and U. Lubeseder. Kennzeichnung gesättigter Formationen. *Math. Zeitschr.*, 82:198–199, 1963.

[46] D. Gorenstein. *Finite Groups*. Chelsea Publishing Company, New York, 1980.

[47] F. Gross. *Überauflösbare Gruppen mit elementar-abelschen Sylowgruppen*. PhD thesis, Christian-Albrechts-Universität, Kiel, 1962.

[48] W. Guo, K. P. Shum, and A. N. Skiba. X-permutable maximal subgroups of Sylow subgroups of finite groups. *Ukrainian Math. J.*, 58:1471–1480, 2006.

[49] W. Guo, K. P. Shum, and A. N. Skiba. Schur-Zassenhaus theorem for X-permutable subgroups. *Algebra Colloq.*, 15:185–192, 2008.

[50] W. Guo and A. N. Skiba. New criterions of existence and conjugacy of Hall subgroups of finite groups. *Proc. Am. Math. Soc.*, 139:2327–2336, 2011.

[51] P. Hall. A note on soluble groups. *J. London Math. Soc (2)*, 3:98–105, 1928.

[52] P. Hall. A characteristic property of soluble groups. *J. London Math. Soc.*, 12:198–200, 1937.

[53] P. Hall. Complemented groups. *J. London Math. Soc.*, 12:201–204, 1937.

[54] P. Hall. On the sylow systems of a soluble group. *Proc. London Math. Soc. (2)*, 43:316–323, 1938.

[55] P. Hall. On the system normalizers of a soluble group. *Proc. London Math. Soc.*, 43:507–528, 1938.

[56] P. Hall. The construction of soluble groups. *J. Reine Angew. Math.*, 182:206–214, 1940.

[57] P. Hall and G. Higman. On the *p*-length of *p*-soluble groups and reduction theorems for burnsides. *Proc. London Math. Soc.*, 6:1–40, 1956.

[58] T. O. Hawkes. Two applications of twisted wreath products to finite soluble groups. *Trans. Am. Math. Soc.*, 214:324–335, 1975.

[59] D. G. Higman. Remarks on splitting extensions. *Pacific J. Math.*, 4:545–555, 1954.

[60] G. Higman. Complementation of abelian normal subgroups. *Publ. Math. Debrecen*, 4: 455–458, 1956.

[61] M. Hofmann. The normal complemented formation. *Commun. Algebra*, 23:5499–5501, 1995.
[62] O. Hölder. Die gruppen der ordnungen p^3, pq^2, pqr, p^4. *Math. Ann.*, 43:301–412, 1893.
[63] O. Hölder. Bildung zusammengesetzter gruppen. *Math. Ann.*, 46:321–422, 1895.
[64] B. Huppert. Gruppen mit modularer Sylow-Gruppe. *Math. Z.*, 75:140–153, 1961.
[65] B. Huppert. Subnormale Untergruppen und p-Sylowgruppen. *Acta Sci. Math. Szeged*, 22: 46–61, 1961.
[66] B. Huppert. *Endlichen Gruppen I*. Springer-Verlag, Berlin, 1983.
[67] I. M. Isaacs. *Finite group theory*. American Mathematical Society, Providence, RI, 2008.
[68] C. Juan-Martánez. *Formaciones Localmente Construibles De Grupos Finitos*. PhD thesis, Universitate de València, València, 1994.
[69] L.-C. Kappe and J. Kirtland. Supplementation in groups. *Glasgow Math. J.*, 42:37–50, 2000.
[70] A. Kertész. On groups every subgroup of which is a direct summand. *Publ. Math. Debrecen*, 2:74–75, 1951.
[71] J. Kirtland. Direct products of inseparable finite groups. *Arch. Math. (Basel)*, 62:289–291, 1994.
[72] J. Kirtland. The geometry of finite groups through the use of splitting systems. *Commun. Algebra*, 22:735–764, 1994.
[73] J. Kirtland. On two classes of finite inseparable p-groups. *Acta Math. Sin. (Engl. Ser.)*, 31: 1203–1214, 2015.
[74] M. Kotzen. On the structure of a finite solvable $\mathcal{K}$-group. *Proc. Am. Math. Soc.*, 27:16–18, 1971.
[75] A. G. Kurosh. The Jordan-Hölder theorem in arbitrary lattices. *Sbornik Pamyati Akad. Grave*, pages 343–346, 1940.
[76] L. J. Leifman and B. F. Johnson, editors. *The Kourovka notebook: unsolved problems in group theory*, volume 121 of *2*. American Mathematical Society Translations, seventh augmented edition, 1983.
[77] B. Li and Z. Zhang. On X-permutable subgroups. *Algebra Colloq.*, 19:699–706, 2012.
[78] U. Lubeseder. *Formationsbildungen in endlichen auflösbaren Gruppen*. PhD thesis, Universität Kiel, 1963.
[79] A. Lucchini, F. Menegazzo, and M. Morgi. On the existence of a complement for a finite simple group in its automorphism group. *Illinois J. Math.*, 47:395–418, 2003.
[80] E. Maillet. Sur les groupes échangeables et les groupes décomposables. *Bull. Soc. Math. France*, 28:7–16, 1900.
[81] G. A. Miller. In a simple group of an odd composite order every system of conjugate operators or sub-groups includes more than fifty. *Proc. London Math. Soc. (1)*, 33:6–9, 1900.
[82] M. F. Newman. On a class of metabelian groups. *Proc. London Math. Soc.*, 10:354–364, 1960.
[83] M. F. Newman. On a class of nilpotent groups. *Proc. London Math. Soc. (3)*, 10:365–375, 1960.
[84] O. Ore. Structures and group theory. I. *Duke Math. J.*, 3:149–174, 1937.
[85] O. Ore. Structures and group theory. II. *Duke Math. J.*, 4:247–269, 1938.
[86] O. Ore. Contributions to the theory of groups of finite order. *Duke Math. J.*, 5:431–460, 1939.
[87] D. J. S. Robinson. *A course in the theory of groups*. Springer-Verlag, New York, second edition, 1996.
[88] J. S. Rose. On a splitting theorem of Gaschütz. *Proc. Edinburgh Math. Soc.*, 15:57–60, 1966.
[89] A. Scarselli. On a class of inseparable finite groups. *J. Algebra*, 58:94–99, 1979.
[90] A. Scarselli. On a class of inseparable finite groups ii. *Accad. Naz. Lincei Rend. Cl. Sci. Fis. Mat. Natur. (8)*, 68:22–25, 1980.
[91] E. Schenkman. The splitting of certain solvable groups. *Proc. Am. Math. Soc.*, 6:286–290, 1955.
[92] R. Schmidt. *Subgroup lattices of groups*. Walter de Gruyter, Berlin, 1994.

[93] O. Schreier. Über die erweiterung von gruppen ii. *Abh. Math. Sem. Univ. Hamburg*, 4:321–346, 1925.

[94] O. Schreier. Über die erweiterung von gruppen i. *Monatsh. Math. Phys.*, 34:165–180, 1926.

[95] I. Schur. Neuer beweis eines satzes über endliche gruppen. *Proc. Sympos. Appl. Math., Sitzber. Akad. Wiss., Berlin*, pages 1013–1019, 1902.

[96] W. R. Scott. *Group Theory*. Dover Publications, New York, 1987.

[97] G. Seitz and C. R. B. Wright. On complements of $\mathfrak{F}$-residuals in finite solvable groups. *Arch. Math.*, 21:139–150, 1970.

[98] L. Šemetkov. Factorization in finite groups. *Sov. Math. Dokl.*, 9:165–168, 1968.

[99] L. Šemetkov. On the existence of π-complements for normal subgroups of finite groups. *Sov. Math. Dokl.*, 11:1436–1438, 1970.

[100] L. Šemetkov. Complements and supplements to normal subgroups in finite groups. *Ukrainian Math. J.*, 23:556–565, 1971.

[101] E. Shult. A note on splitting in solvable groups. *Proc. Am. Math. Soc.*, 17:318–320, 1966.

[102] M. Suzuki. *Structure of a Groups and the Structure of Its Lattice of Subgroups*. Springer-Verlag, Berlin, 1956.

[103] M. Suzuki. *Group Theory I*. Springer-Verlag, Berlin, 1982.

[104] D. R. Taunt. On *A*-groups. *Proc. Cambridge Philos. Soc.*, 45:24–42, 1949.

[105] J. Thompson. Finite groups with fixed-point-free automorphisms of prime order. *Proc. Natl. Acad. Sci. U.S.A.*, 45:578–581, 1959.

[106] M. Weinstein, editor. *Between Nilpotent and Sovable*. Polygonal Publishing House, Passaic, NJ, 1982.

[107] L. Weisner. A theorem concerning direct products. *Bull. Am. Math. Soc.*, 33:44–45, 1927.

[108] J. Wiegold. On direct factors in groups. *J. London Math. Soc.*, 35:310–319, 1960.

[109] H. Wielandt. Entwicklungslinien in der Strukturtheorie der Endlichen gruppen. In *Proceedings of the International Congress of Mathematicians (Edinburgh 1958)*, pages 268–278, New York, 1960. Cambridge University Press.

[110] C. R. B. Wright. On screens and $\mathfrak{L}$-izers of finite solvable groups. *Math. Z.*, 115:273–282, 1970.

[111] C. R. B. Wright. On complements to normal subgroups in finite solvable groups. *Arch. Math.*, 23:125–132, 1972.

[112] M. Yonaha. A splitting condition for metabelian groups. *Bull. Arts Sci. Div. Univ. Ryukyus Math. Nat. Sci.*, 7:1–4, 1964.

[113] G. Zacher. Caratterizzazione dei gruppi risolubili d'ordine finito complementati. *Rend. Sem. Mat. Univ. Padova*, 22:113–122, 1953.

[114] H. Zassenhaus. *Lehrbuch der Gruppentheorie*. Hamburg Mathematische Einzelschriften, No. 21. Teubner, Leipzig and Berlin, 1937.

Author index

DOI 10.1515/9783110480214-010

Subject index

DOI 10.1515/9783110480214-011

Bei Fragen zur Produktsicherheit wenden Sie sich bitte an:
If you have any questions regarding product safety,
please contact:

Walter de Gruyter GmbH
Genthiner Straße 13
10785 Berlin
productsafety@degruyterbrill.com